Controlling Language in Industry

Stephen Crabbe

Controlling Language in Industry

Controlled Languages for Technical Documents

Stephen Crabbe
University of Portsmouth
Portsmouth, United Kingdom

ISBN 978-3-319-52744-4   ISBN 978-3-319-52745-1 (eBook)
DOI 10.1007/978-3-319-52745-1

Library of Congress Control Number: 2017930262

© The Editor(s) (if applicable) and The Author(s) 2017
This work is subject to copyright. All rights are solely and exclusively licensed by the Publisher, whether the whole or part of the material is concerned, specifically the rights of translation, reprinting, reuse of illustrations, recitation, broadcasting, reproduction on microfilms or in any other physical way, and transmission or information storage and retrieval, electronic adaptation, computer software, or by similar or dissimilar methodology now known or hereafter developed.
The use of general descriptive names, registered names, trademarks, service marks, etc. in this publication does not imply, even in the absence of a specific statement, that such names are exempt from the relevant protective laws and regulations and therefore free for general use.
The publisher, the authors and the editors are safe to assume that the advice and information in this book are believed to be true and accurate at the date of publication. Neither the publisher nor the authors or the editors give a warranty, express or implied, with respect to the material contained herein or for any errors or omissions that may have been made.

Cover illustration: Détail de la Tour Eiffel © nemesis2207/Fotolia.co.uk

Printed on acid-free paper

This Palgrave Macmillan imprint is published by Springer Nature
The registered company is Springer International Publishing AG
The registered company address is: Gewerbestrasse 11, 6330 Cham, Switzerland

For Samantha, Lachlan and Brownie

Acknowledgements

I am grateful to Terry Bell, Dr. Ralph Calistro, John Smart and Anne Zuza for providing me with difficult-to-access material both from and about existing controlled languages for technical documents.

An earlier version of parts of Chap. 1 was previously published in article form in:

Crabbe, Stephen J. 2012. Constructing a Contextual History of English Language Technical Writing. *Journal of Translation and Technical Communication Research* 5(1): 40–59.

Contents

1 Introduction and Historical Development of Technical Documents   1

2 Existing Controlled Languages for Technical Documents   23

3 Best-Practice Features of Modern Technical Documents   49

4 Analysing Existing Controlled Languages Against the Best-Practice Features   69

5 Developing a New Controlled Language for Technical Documents   89

6 Trialling a New Controlled Language for Technical Documents   107

Index   115

List of Abbreviations

BCE   Bull Controlled English
CFE   Caterpillar Fundamental English
CTE   Caterpillar Technical English
EE   Ericsson English
NSE   Nortel Standard English
PACE Perkins Approved Clear English

List of Figures

Fig. 2.1	Categories of controlled language rules	26
Fig. 2.2	Comparison of the BCE and CFE rules	41
Fig. 3.1	Linguistic best-practice features	51
Fig. 3.2	Organisational best-practice features	57
Fig. 4.1	Analysed best-practice features	74
Fig. 4.2	Breakdown of the identified rules	84
Fig. 4.3	Breakdown of the identified best-practice features	85
Fig. 5.1	Development of rules 1.1–1.7	94
Fig. 5.2	Development of rules 2.1–2.5	95
Fig. 5.3	Development of rules 3.1–3.7	96
Fig. 5.4	Development of rules 4.1–4.7	98
Fig. 5.5	Development of rules 5.1–5.3	99
Fig. 5.6	Examples of some rules in practice	100
Fig. 5.7	Extract from an example word frequency list	102
Fig. 6.1	Balance of question types	111

1

Introduction and Historical Development of Technical Documents

Abstract The chapter introduces the book, its aim and structure. It then sets the book's context by outlining the historical development of English language technical documents from the pre-industrial fourteenth century to the industrial early twentieth century and the introduction of Basic English, a precursor to modern controlled languages for technical documents. In doing this, the chapter identifies a number of linguistic and organisational features of historical technical documents that are used in modern technical documents to help make the information in them understandable.

Keywords American industrial revolution • British industrial revolution • German industrial revolution • Historical technical documents • Technical document features

Introduction

Controlled languages are languages with a simplified set of rules and controlled vocabulary that shape and constrain the information in technical documents to help make it understandable and, in many cases, aid

© The Author(s) 2017
S. Crabbe, *Controlling Language in Industry*,
DOI 10.1007/978-3-319-52745-1_1

machine translation. Since the 1970s, many large manufacturing companies have developed controlled languages for their technical documents. However, a lack of detailed information about controlled languages persists to this day as many existing ones are in-house language systems with little information or research publicly disseminated about them. As a result, there is currently no complete book on the market that addresses controlled languages. This means that manufacturing companies or individuals unfamiliar with controlled languages have to identify and collate information from multiple sources, making it challenging to understand and develop them.

The aim of this book is to help fill this gap through introducing readers to the purpose, use, development and trialling of controlled languages. The main objective in doing this is to provide readers with a clear understanding of existing controlled languages for technical documents and models which they can use or adapt to develop and trial a new controlled language for their own technical documents.

The book is divided into six chapters. Chapter 1, the current chapter, introduces the book and sets its context by outlining the historical development of technical documents from the pre-industrial fourteenth century to the industrial early twentieth century and the introduction of Basic English, a precursor to modern controlled languages for technical documents. In doing this, the chapter shows the historical nature of some of the linguistic and organisational best-practice features of modern technical documents that help make the information in them understandable.

Chapter 2 takes Basic English as a starting point from which to provide a chronological description of controlled languages that have been used by large manufacturing companies for their technical documents since the 1970s, particularly to help make the information in them understandable. Specifically, it describes why and how each controlled language was developed and used. The chapter concludes by identifying the benefits and drawbacks of controlled languages.

Chapter 3 brings together and summarises the most widely agreed upon linguistic and organisational best-practice features of modern technical documents that help make the information in them understandable. It concludes by presenting the key benefits of developing understandable technical documents.

Chapter 4 compares the controlled language rule sets introduced in Chap. 2 with the linguistic and organisational best-practice features of modern technical documents introduced in Chap. 3 to ascertain whether, and to what extent, the rules reflect the best-practice features. The findings from the analysis help in identifying new non-manufacturer specific rules for a controlled language rule set model.

Chapter 5 draws from all the previous chapters to provide models for developing a controlled language rule set and controlled vocabulary that readers can use or adapt to develop a new controlled language for their technical documents.

Chapter 6 concludes the book by providing a model for trialling a new controlled language that readers can use when deciding whether or not to implement a new controlled language fully.

Historical Development of Technical Documents: An Overview

In 1985, Moran argued that the history of technical writing had not yet been written. This is not to suggest that there was no literature on English language technical writing at this time. However, Hills and McLaren (1987) point out that much of the limited available literature blended the histories of scientific and technical writing so as to integrate the large and well-developed body of knowledge about scientific writing with the small and less-developed body of knowledge about technical writing. Researchers such as Moran and Journet (1985) and Hills and McLaren (1987) argued for distinguishing between scientific and technical writing so as to allow technical writing to develop, among other things, its own distinct history and field of research. However, despite these arguments, Tebeaux and Killingsworth (1992) make clear that technical writing still did not have its own separate history and identity by the early 1990s.

Several publications took a first step towards addressing this issue in the late 1990s, including book-length works by Tebeaux (1997),

Brockmann (1998) and Kynell and Moran (1999). In a 1999 bibliographical essay, Rivers drew attention to this growing interest in technical writing but argued for more contextual approaches to historical studies in the field. And, in 2000, the field did benefit from two richly contextual, book-length historical studies on technical writing by Longo (2000) and Kynell (2000).

In 2002, Kynell and Seely emphasised the need for further contextual historical studies. However, relevant historical studies remained limited. In 2003, Savage thus wrote that the history of technical writing is "in the very early stages of being written" (3–4). This view has been echoed by Tebeaux (2008) and Moran and Tebeaux (2011). Tebeaux has recently started to address it with the publication of *The Flowering of a Tradition: Technical Writing in England 1641–1700* (2014). However, the book begins by drawing attention to the limited historical literature available:

> This monograph marks my second effort to present a sustained description of the history of English technical writing. *The Emergence of a Tradition*, which covered 1475–1640, the heart of the English Renaissance, still stands as the only effort to describe, in broad outline, examples of technical writing in that period. (Tebeaux 2014: v)

The rest of this chapter thus not only sets the book's context but also contributes to addressing the need for contextual historical descriptions of technical writing. It does this through outlining the historical development of English language technical documents from the fourteenth to early twentieth century and, in doing so, identifying a number of linguistic and organisational features of historical technical documents that will be seen in Chap. 3 as being used in modern technical documents to help make the information in them understandable.

Pre-industrial Britain

Before the beginning of the industrial revolution around the second half of the eighteenth century, Britain was overwhelmingly rural-based and agrarian-dependent. Production, trade and consumption were con-

1 Introduction and Historical Development of Technical... 5

sequently focused on agriculture. However, this does not mean that there was not demand for, and production of, non-agrarian goods such as candles, clothing, household furniture, kitchen utensils and shoes. The majority of these non-agrarian goods were produced in family-based rural cottage industries and urban-based craft and merchant guild workshops.

The production of textiles for clothing was a typical family-based rural cottage industry. The production unit was primarily the family, with the father responsible for weaving and apprenticing male offspring into weaving and the mother responsible for spinning and apprenticing female offspring into spinning.

The spinning wheel was the most widely used machine of the pre-industrial textile cottage industry. This type of early technology required human operation. However, the operation was simple enough that knowledge of it successfully passed intergenerationally from father to male offspring, or mother to female offspring, through observation and imitation.

Technical knowledge diffusion is widely associated with technical documents in post-industrial revolution societies. However, the limited complexity, and the relatively unchanged and unchanging nature, of much of the technology that was used in the pre-industrial cottage industries and workshops of Britain meant that technical knowledge diffusion was largely visual and verbal. This does not mean that there were no technical documents. In fact, early precursors of modern technical documents can be traced as far back as the fourteenth century. One example is Chaucer's *A Treatise on the Astrolabe* (1391).

The word astrolabe comes from the Greek words ἄστρο (astro) meaning star and λάβιον (labio) meaning finder. The astrolabe is a technical instrument that was commonly used in pre-industrial societies to find the position of the stars in order to determine time and latitude. The purpose and structure of Chaucer's *A Treatise on the Astrolabe* are introduced in the prologue:

> I purpose to teche thee a certayn nombre of conclusions pertaynyng to this same instrument ... The firste partye of this tretyse shal shewe the figures and the membres of thyn astrolabe, bicause that thou shalte have the greter

knowinge of thyne owne instrument ... The seconde partye shal teche thee to werken the verray practike of the forsayde conclusions. ([1391] 1870: 19)

Its purpose is thus instructional. It is divided into two sections with the first section providing a description of the parts of the astrolabe and the second section providing instructions on its operation. The first section is ten pages and the second section is 41 pages, so most of it is given over to providing instructions on how to operate this early technology.

The information in any technical document should be understandable for the target reader. Chaucer's target reader was a ten-year-old child who is referred to as lytel Lowys and he employs several methods to help lytel Lowys understand the information. First, he writes in English rather than Latin. To give this some context, English was regarded at this time as one of the "tongues of the common folk" (Hager and Nelson 1993: 88). He also uses a simplified grammar and lexicon or "lighte reules and nakyd words in englissh" ([1391] 1870: 19). As a result, Basquin argues that the information in *A Treatise on the Astrolabe* is still understandable to this day "with only a little effort by any educated reader of English" (1981: 22).

The information in the second section is organised into 42 short, task-oriented chunks. Each chunk provides instructions on how to operate the astrolabe for a specific task, as can be seen in the following extract from two of the chunks (II and XVIII).

> II. To knowe the altitude of the sonne, eyther of celestiale bodies.
> Sette the ryng of thyne astrolabie upon thy ryghte thombe and tourne thy lyfte syde again the light of the sonne and remeve thy rewle up and downe till the streme of the sonne shyne through bothe holes of the rewle.
> XVIII. To knowe the degre of the sonne by the rete for a maner curyosyte.
> Seke busely with thy rule the highest of the sonne in myddes of the daye; tourne than thyn astrolabie, and with a prycke of ynk marke the nombre of the same altitude in the lyne meridionale. ([1391] 1870: 33 and 42)

The extract shows, first, that the headings "To knowe the altitude of the sonne, eyther of celestiale bodies." and "To knowe the degre of the sonne by the rete for a maner curyosyte." are task-oriented and have a parallel language structure and, second, that the corresponding instructional sentences are in the active voice and imperative mood. As will be seen in Chap. 3, these are linguistic and organisational best-practice features of modern technical documents that are used to help make the information in them understandable. Historically, they can also be found in Chaucer's fourteenth-century technical document.

It is interesting to briefly note that *A Treatise on the Astrolabe* may not only have been used to provide information on the operation of the astrolabe. Eagleton argues that several astrolabes were constructed to the same design as the one in Chaucer's technical document. This suggests that the information in it was used to construct them. Eagleton concludes that "Chaucer's name on a book could sell it, so also a Chaucerian astrolabe would have held a certain prestige" (2007: 324).

English language technical writing started to increase in quantity and diversity during the renaissance period of the sixteenth and seventeenth centuries to include technical dictionaries such as Harris' *Lexicon Technicum* or *An Universal English Dictionary of Arts and Sciences* and Chambers' *Cyclopedia* or *An Universal Dictionary of Arts and Sciences*.

Harris' technical dictionary was first published in 1704. It was the first technical dictionary in the English language, and researchers such as Russell (1997), Mather and Solberg (2000) and Yeo (2001) refer to Harris with the epithet Technical Harris due to the large number of descriptions and illustrations of pre-industrial technology in this dictionary.

Chambers' technical dictionary was first published in 1728. It contains a wide range of entries under the heading mechanical arts. Pannabecker (1994) points out that the terms mechanick trades and mechanical arts were more frequently used than the term technology in the eighteenth century. Chambers describes the entries under the heading mechanical arts as "being all practised by means of some Machine or Instrument" (1728: 144). The dictionary entry for a mechanical printing press is typical in its structure: providing an overview of this early technology, a description of its parts and then a description of its operation. The following extract is taken from the description of its parts.

Under the Carriage is fix'd a small piece of Iron call'd the *Spit*, with a double Wheel in the middle, round which Leather *Girts* are fasten'd, nail'd to each end of the *Plank*. To the outside of the *Spit* is fix'd a Handle, or *Rounce*, by which the Press-man turns the Plank in or out at pleasure. (Chambers 1728: 877)

Chambers' description of the way that the "*Spit*" and the "*Plank*" fit together, the operation of the "*Spit*" to move the "*Plank*" and the technical terms "*Spit*" and "*Rounce*" is still understandable to the modern reader. Yet, Chambers was writing at a time when, as pointed out by Wasson, "written English was often ponderous, wordy, and convoluted" (2009: 24).

The *Lexicon Technicum* and *Cyclopedia* were both published by subscription; thus, their target readers can be ascertained from their subscription lists. The 1704 edition of the *Lexicon Technicum* has 903 listed subscribers and the 1728 edition of *Cyclopedia* has 375 listed subscribers. These subscribers are dominated by British royalty, aristocracy and wealthy men of letters. For example, the 375 subscribers listed in the *Cyclopedia* include 42 Privy Councillors with the title "The Right Hon." or "The Rt. Hon.", 32 members of the clergy with the title "The Rev." or "The Reverend" and 32 Knights with the title "Sir" (Chambers 1728: 3–4). Harris and Chambers were thus both writing for a numerically small British elite on whom they depended for money. As a result, their works were unlikely to have been read by those members of the labouring population operating the pre-industrial technology described in them or to have extended much beyond Britain.

French was, in fact, the pan-European lingua franca of intellectual discourse during this period; thus, French language technical writing was more likely to have had a larger and broader readership. The most comprehensive technical dictionary of the pre-industrial period was not the English language *Lexicon Technicum* or *Cyclopedia*; rather it was the French language *Encyclopédie, ou Dictionnaire raisonné des sciences, des arts et des métiers* (1751) by Diderot and D'Alembert. This provided detailed descriptions and illustrations of the tools, instruments, machines and techniques that were used in eighteenth-century French cottage industries and workshops. However, it should be noted that while Diderot and

D'Alembert did extensively document the eighteenth-century mechanick trades or mechanical arts in France, their readers also did not necessarily include the people in the cottage industries and workshops using the machines. Darnton argues that "although one cannot exclude the possibility that the *Encyclopédie* reached a great many readers in the lower classes, its main appeal was the traditional elite" (1986: 299).

Written and printed forms of technical knowledge were thus produced for a limited readership during the pre-industrial period. However, this began to change around the middle of the eighteenth century with the onset of the British industrial revolution and the transition from family-based to factory-based production and hand-driven to technology-driven machines.

British Industrial Revolution

The British textile industry led the way in this transition, with the second half of the eighteenth century marked by a succession of mechanical inventions such as Arkwright's water frame (1769), Hargreaves' spinning jenny (1769), Crompton's spinning mule (1779), Cartwright's power loom (1785) and Whitney's cotton gin (1793). These machines were too expensive for individual families and too large for individual family homes. There was a resultant multiplication of factories to accommodate them during the last third of the eighteenth century.

In 1829, Carlyle lamented that he was living in the "Age of Machinery … the living artisan is driven from the workshop to make room for a speedier, inanimate one. The shuttle drops from the hand of the weaver, and falls into iron ones that ply it faster … There is no end to machinery" (1829: 34). However, adoption of this new technology, and centralisation of production, was a gradual process with marked differences between industries. Furthermore, new technology could be invented, but workers with experience of constructing, operating and maintaining it did not exist. As a result, the hitherto successful method of orally passing on technical knowledge from generation to generation became increasingly less relevant.

Reference to Tebeaux reveals that "the emergence of technical writing is the story of the shift from orality to textuality; and technical writing echoes the literacy of both the writer and the intended readers" (2008: 20). Overall, literacy levels increased during the industrial revolution. These increases were particularly marked in the industrialising cities due to the range and diversity of educational establishments that became available to the growing factory-based labouring classes.

The Mechanics' Institute was one of the most prominent of these educational establishments. The first institute was established in Glasgow in 1821, and, by 1850, it is estimated by researchers (for example, Woodward 1963; Auerbach 1999) that there were more than 600 institutes. Many famous names from the industrial revolution—such as Robert Stephenson, Joseph Locke and James Hall Nasmyth—gave lectures on technical subjects at institutes. In addition, many leading manufacturing companies from the industrial revolution—such as the Eastern Counties Railway, George Stephenson and Company and the Royal Arsenal—established new institutes. However, the reason for much of this technical knowledge dissemination was not necessarily altruistic, merely practical. Workers required training on how to operate the machines not for their protection but for the protection of the machines. Crompton's spinning mule, for example, was a very expensive and complicated machine, and von Tunzelmann (1994) suggests that it took around three months for an experienced spinner to learn how to operate one.

Many of the institutes offered literacy classes. Furthermore, the sixth clause of the Health and Morals of Apprentices Act of 1802 made it a legal requirement for every factory apprentice to be given reading and writing classes. Court (1954) points out that having workers with basic literacy was important in the industrialising cities as technical training was difficult to carry out without it. Manufacturing companies also needed workers who could construct and maintain the new machines. Mokyr (2006) describes these workers as tens of thousands of literate mechanics and craftsmen who were able to understand technical writing and illustrations.

It is interesting to note that the geographical focus of these developments in technology and literacy in Britain meant that the English language contributed to the dissemination of technical knowledge across continental Europe. This is not to say that it was actually intended thus. The British government, in fact, tried to stop the spread of British technology and technical documents to continental Europe in successive acts of 1750, 1774, 1781 and 1785. The 1781 act, for example, made it illegal to export "any machine, engine, tool, press, paper, utensil, or implement …. any part or parts of such machine, engine, tool, press, paper, utensil, or implement … or any model or plan, or models or plans, of any such machine, engine, tool, press, paper, utensil, or implement" (Evans 1836: 178). It also tried to limit the emigration of British mechanics and craftsmen through a 1718 act titled *An act to prevent the inconvenience arising from seducing Artificers in the Manufactures of Great Britain into foreign parts.*

Continental European manufacturing companies did, however, acquire British technology, British technical documents and literate British mechanics and craftsmen, as is illustrated in the following extract from an 1824 government select committee interview with Martineau, a leading British industrialist during the industrial revolution:

> Cannot specifications of every new machine, with drawings and models, be obtained from this country?
> I have no doubt they may be, and are daily procured; I was conversing last week with a large cotton manufacturer from France, who stated distinctly, that there was no model or machine in existence in England, of which he could not obtain a model or drawing by paying for it.
> Are those models, drawings, and specifications such, that when carried abroad, they may by expert English artizans be made up and completed? Certainly they may be. (House of Commons 1824: 7)

Reference to Crystal (2003), in fact, reveals that the industrial revolution witnessed a huge growth in technology-related publications. The increasing literacy of the labouring classes, and their increasing exposure to new technology, created an expanding market for technology-related books and magazines. Examples of the technology-related books include *Observations on the Use of Power Looms by a Friend to the Poor* (Friend to the Poor 1823), *The New Invention of Double and Quadruple or British National Looms* (Sadler 1831) and *Scott's Practical Cotton Spinner* (Scott

1851). Examples of the technology-related magazines include *Glasgow Practical Mechanics' and Engineers' Magazine*, *Practical Mechanics Journal* and *Mechanics' Magazine, Museum, Register, Journal, and Gazette*.

The French industrialist Motte-Bossut described Britain at the end of the industrial revolution as "the centre of the most advanced industry of Europe and of the Universe" (cited by Stearns 2007: 53). However, manufacture was also beginning to expand and mechanise in other countries, particularly the USA and Germany.

American Industrial Revolution

Stevens (1995) suggests that learning how to operate machines in the USA during the late eighteenth and nineteenth century required access to previously constructed ones. However, this period saw the publication of, in the words of Ong, "how-to-do-it manuals for the trades" (2002: 43), an example of which is Evans' *The Young Mill-Wright and Millers Guide*. This was first published in 1795 and reprinted 14 times in English by 1860. Reference to Kynell and Moran (1999) reveals that it was the most reprinted technical publication in the USA prior to 1861. Its target readers were aspiring mill-wrights and millers, and its purpose was to provide them with all the information necessary to construct, operate and maintain the milling machine that Evans had designed without necessarily having access to a previously constructed one.

Evans' technical document is divided into five sections, with the final section containing the instructions on constructing, operating and maintaining the milling machine. This information is organised into short, task-oriented chunks as with Chaucer's *A Treatise on the Astrolabe*. The following extract is taken from one of the chunks.

> Directions for constructing undershot wheels, such as shown in figure I, plate XIII.
> 1. Dress the arms straight and square on all sides, and find the centre of each; divide each into 4 equal parts on the side, square, centre, scribe and gauge them from the upper side across each point, on both sides, 6 inches each way from the centre.

1 Introduction and Historical Development of Technical... 13

2. Set up a truckle or centre post, for a centre to frame the wheel on, in a level piece of ground, and set a stake to keep up each end of the arms level with the truckle, of convenient height to work on.
3. Lay the first arm with its centre on the centre of the truckle, and take a square notch out of the upper side 3–4ths of its depth, wide enough to receive the 2d arm.
4. Make a square notch in the lower edge of the 2d arm, 1–4th of its depth, and lay it in the other, and they will joint standing square across each other. (Evans 1850: 304)

This reveals several recurring features of Evans' technical document. First, the heading is task-oriented. Second, the instructional sentences are in the active voice and imperative mood. Third, numbers are used to indicate the consecutive nature of the instructional sentences and group together related instructions. These will be seen in Chap. 3 to be linguistic and organisational best-practice features of modern technical documents that are used to help make the information in them understandable. Historically, they were also being used by Evans in the late eighteenth century.

Tebeaux and Killingsworth define technical writing as "writing that enables readers to perform tasks associated with their work" (1992: 8). This definition is as applicable to Evans' nineteenth-century technical document as it is to modern technical documents. However, a new kind of technical writing also appeared during the second half of the nineteenth century. Manufacturers of sewing machines such as the American Sewing Machine Company, Singer Sewing Machine Company and Wilson Sewing Machine Company started producing domestic sewing machines for private use and brought with them the first technical documents for consumer technology.

The opening pages of this new kind of technical document had two purposes. The first was to persuade the end users of the new consumer technology that operating it was not difficult. For example, the opening page of an 1870 technical document for a Wilson sewing machine states that it is a "paragon of simplicity, and the most unsophisticated persons can learn to use it effectually" (Wilson Sewing Machine Company 1870: 1).

The second purpose was to show the importance of the information in the technical document. The first page of the aforementioned 1870 technical document also states that "You cannot use the Machine until you thoroughly understand the Directions … We have prepared very explicit instructions; with which, any one, from a careful perusal, can learn to use it" (Wilson Sewing Machine Company 1870: 1). These dual purposes were particularly important during the second half of the nineteenth century when consumer technology was still new and unfamiliar. The following extract is taken from the aforementioned 1870 technical document.

> To Set the Needle.
> Place the shank of the needle in the needle bar with the long groove of the needle to the left or outside of the machine, and secure it with the screw at the end of the needle bar, then raise the needle bar to its highest point, and with the left hand insert the point of a fine needle or pin into the eye, resting the same on the throat-plate; then loosen the needle screw, and hold the instrument so that the eye of the needle will be in a direct line from left to right, then turn the balance wheel, up or down as may be required, until the gauge mark on the needle bar rests even with the top part of the face plate, then secure the needle firmly with the needle screw. (Wilson Sewing Machine Company 1870: 4)

This reveals that the heading is again task-oriented and that the instructional sentence is again in the active voice and imperative mood. Numbers are not used. However, the consecutive nature of the instructions is indicated by the repeated use of "then."

Growing sales of domestic sewing machines in the early twentieth century brought technical documents increasingly into the homes of ordinary American consumers, as did another American mass consumer good: Ford's Model T car.

Ford stated that he would produce a car "for the great multitude … it will be so low in price that no man making a good salary will be unable to own one" (Ford 1923: 73). The result was the Model T car. New owners were given a technical document entitled *Ford Manual For Owners and Operators of Ford Cars and Trucks*, the opening pages of which echo the 1870 technical document for a Wilson sewing machine.

> The Ford is the simplest car made. It is easy to understand, and it is not difficult to keep in proper adjustment and repair. That the Ford construction may be thoroughly understood—and that there may be an authoritative guide for the making of Ford adjustments—this book is published. (Ford Motor Company 1919: 2)

The information in this technical document is also organised into short, task-oriented chunks, such as the one shown.

> How is the Car stopped?
> Partially close the throttle; release the high speed by pressing the clutch pedal forward into neutral; apply the foot brake slowly but firmly until the car comes to a dead stop. Do not remove foot from the clutch pedal without first pulling the hand lever back to neutral position, or the engine will stall. To stop the motor, open the throttle a trifle to accelerate the motor and then throw off the switch. The engine will then stop with the cylinders full of explosive gas, which will naturally facilitate starting.
> Endeavor to so familiarize yourself with the operation of the car that to disengage the clutch and apply the brakes becomes practically automatic-the natural thing to do in case of emergency. (Ford Motor Company 1919: 7)

The heading "How is the Car stopped?" is task-oriented and the instructional sentences are in the active voice and imperative mood. It is also interesting to note that the average number of words per instructional sentence is 24. Yet, the number of words in the instructional sentence for setting the needle from the 1870 technical document for the Wilson sewing machine is 135. This suggests a historical movement towards shorter, more simply structured sentences in technical documents.

The early twentieth century also saw the introduction of dedicated books and courses on technical writing. Researchers such as Kynell (2000) and Gianniny (2004) agree that the first book on English language technical writing was Rickard's 1908 publication *A Guide to Technical Writing*. Rickard's advice is equally pertinent to modern technical communicators.

> Conscientious writers try to improve their mode of expression by precision of terms, by careful choice of words, and by the arrangement of them so that they become efficient carriers of thought from one mind to another. (1908: 8)

The second book was Earle's 1911 publication *The Theory and Practice of Technical Writing*. Connors (1999) reveals that not only was this the first dedicated technical writing book to be used in a higher education course, but also that the course taught by Earle at Tufts College in Massachusetts was the first dedicated technical writing course.

The prominence of the USA in the discussion thus far might give the incorrect impression that this expansion in consumer technology and associated technical documents was an overwhelmingly American phenomenon. However, this was most definitely not the case.

German Industrial Revolution

By the early twentieth century, German low-priced consumer technology was flooding Britain. Reference to Head (1992) reveals that the German-produced goods were equal in quality to, if not better than, British ones. The Leica range of cameras from the German manufacturing company Ernst Leitz Gmbh exemplifies this commitment to quality. The Leica first became commercially available in 1925 and is described by White as "a masterpiece of precision craftsmanship" (2001: 64).

The opening page of a 1937 technical document for the setup, operation and maintenance of a Leica camera is similar in language and purpose to the earlier technical documents for the Wilson sewing machine and Ford car.

> For the description of even the simple manipulations called for in the operation of the Leica camera much space may be required, in the form of text matter or illustrations. If at the outset the camera and the instructions for use are taken in hand together, most of the following instructions will at once become obvious. (Ernst Leitz Gmbh 1937: 3)

The following extract shows the instructions for taking a photograph.

> D. Taking the Photograph
> 1. Pull out lens, and turn it to the right (clockwise) so as to lock it in the bayonet catch.

2. Adjust iris diaphragm by means of lever or ring 21 (Fig. 18).
3. Wind knob 1 in direction of arrow right to stop.
4. See that shutter speed is correct or set it by lifting the speed dial 7, at the same time turning it so that the required figure lies against the index arrow 8. Let go knob which will then settle in position. At Z the shutter remains open as long as the button is pressed down.
4a. See further remarks page 22 re Leica Model F and G.
5. Sight the object through range finder 11, turning focusing lever 17 until the two images coincide (fuse into one). Use view-finder 10 to view the whole field and gently (not jer-kily) release press button 5. When photographing rapidly moving objects the range finder should be used as view-finder. (Ernst Leitz Gmbh 1937: 15)

This reveals that the heading is again task-oriented. The instructional sentences are again in the active voice and imperative mood. Furthermore, numbers are used to indicate the consecutive nature of the instructional sentences and group together related instructions. The average number of words per instructional sentence is 15. In comparison, the average number of words per instructional sentence in the extract from the 1919 technical document for the Ford car is 24. This again suggests a gradual historical movement towards shorter, more simply structured sentences in technical documents.

Summary

The main objective of this chapter has been to introduce the book and set its context through tracing the historical development of technical documents from the pre-industrial fourteenth century to the industrial early twentieth century. In doing this, it has identified a number of linguistic and organisational features of historical technical documents—the use of the active voice and imperative mood in instructional sentences, the organisation of instructional sentences into task-oriented chunks and the use of task-oriented headings, parallel language structures and shorter, more simply structured sentences—that will be seen in Chap. 3 as being used in modern technical documents to help make the information in them understandable.

It was against the early twentieth-century background of the ever-increasing internationalisation and technologisation of production that Ogden developed a simplified international language called Basic English. This was not developed specifically for English language technical documents. However, it is adaptable to them. Furthermore, it is widely agreed that many controlled languages developed specifically for English language technical documents are derived from it (for example, Thomas et al. 1992; Kaji 1999; Hartley and Paris 2001; Weiss 2005). A complete understanding of these controlled languages thus necessitates first looking at Basic English in the next chapter.

Bibliography

Auerbach, Jeffrey A. 1999. *The Great Exhibition of 1851: A Nation on Display.* New Haven: Yale University Press.
Basquin, Edmond A. 1981. The First Technical Writer in English: Geoffrey Chaucer. *Technical Communication* Third Quarter: 22–24.
Brockmann, R. John. 1998. *From Millwrights to Shipwrights to the Twenty-First Century.* Cresskill: Hampton Press.
Carlyle, Thomas. 1829. Sign of the Times. In *The Spirit of the Age: Victorian Essays*, ed. Gertrude Himmelfarb, 31–49. New Haven: Yale University Press.
Chambers, Ephraim. 1728. *Cyclopaedia; or, An Universal Dictionary of Arts and Sciences.* London: W. Innys, A. Ward, J. and P. Knapton, T. Osborn, S. Birt, D. Brown et al.
Chaucer, Geoffrey. [1391] 1870. *A Treatise on the Astrolabe.* London: John Russell Smith.
Connors, Robert J. 1999. The Rise of Technical Writing Instruction in America. In *Three Keys to the Past: The History of Technical Communication*, ed. Teresa C. Kynell and Michael G. Moran, 173–196. Westport: Greenwood Publishing Group.
Court, William H.B. 1954. *A Concise Economic History of Britain: From 1750 to Recent Times.* Cambridge: Cambridge University Press.
Crystal, David. 2003. *English as a Global Language.* 2nd ed. Cambridge: Cambridge University Press.
Darnton, Robert. 1986. *The Business of Enlightenment: A Publishing History of the Encyclopédie 1775–1800.* Cambridge: Harvard University Press.

Diderot, Denis, and Jean Le Rond D'Alembert. 1751. *Encyclopédie, ou dictionnaire raisonné des sciences, des arts et des metiers*. Geneve: Chez Pellet, Imprimeur-Libraire.
Eagleton, Catherine. 2007. 'Chaucer's Own Astrolabe': Text, Image and Object. *Studies in History and Philosophy of Science* 38(2): 303–326.
Earle, Samuel C. 1911. *The Theory and Practice of Technical Writing*. New York: The Macmillan Company.
Ernst Leitz Gmbh. 1937. *Leitz Directions Leica Camera*. Germany: Ernst Leitz Gmbh.
Evans, William D. 1836. *A Collection of Statutes Connected with the General Administration of the Law According to the Order of the Subjects*. Vol. 4. 3rd ed. London: W.H. Bond.
Evans, Oliver. 1850. *The Young Mill-Wright and Miller's Guide*. 13th ed. Philadelphia: Lea & Blanchard.
Ford, Henry. 1923. *My Life and Work*. Garden City: Doubleday, Page & Company.
Ford Motor Company. 1919. *Ford Manual for Owners and Operators of Ford Cars and Trucks*. Detroit: Ford Motor Company.
Gianniny, Omer A. 2004. A Century of ASEE and Liberal Education (or How Did We Get from There, and Where Does It All Lead?). In *Liberal Education in Twenty-First Century Engineering: Responses to ABET/EC 2000 Criteria*, ed. David F. Ollis, Kathryn A. Neeley, and Heinz C. Luegenbiehl, 320–346. New York: Peter Lang Publishing.
Hager, Peter J., and Ronald J. Nelson. 1993. Chaucer's 'A Treatise on the Astrolabe': A 600-year-old model for Humanizing Technical Documents. *IEEE Transactions on Professional Communication 36*(2): 87–94.
Harris, John. [1704] 2006. *Lexicon Technicum; or, An Universal English Dictionary of Arts and Sciences*. Mansfield Centre: Martino Publishing.
Hartley, Anthony, and Cécile Paris. 2001. Translation, Controlled Languages, Generation. In *Exploring Translation and Multilingual Text Production: Beyond Context*, ed. Erich Steiner and Colin Yallop, 307–326. Berlin: Walter de Gruyter.
Head, David. 1992. *Made in Germany: The Corporate Identity of a Nation*. London: Hodder & Stoughton.
Hills, Philip, and Margaret McLaren. 1987. *Communication Skills: An International Review*. London: Croom Helm.
House of Commons. 1824. *First Report from Select Committee on Artizans and Machinery*. London: House of Commons.

Kaji, Hiroyuki. 1999. Controlled Languages for Machine Translation: State of the Art. In *Proceedings of the Machine Translation Summit VII*, ed. Miriam Butt and Tracy H. King, 37–39. Singapore: The Asia-Pacific Association for Machine Translation.

Kynell, Teresa C. 2000. *Writing in a Milieu of Utility: The Move to Technical Communication in American Engineering Programs, 1850–1950*. Westport: Greenwood Publishing Group.

Kynell, Teresa C., and Michael G. Moran. 1999. *Three Keys to the Past: The History of Technical Communication*. Stamford: Ablex Publishing Corporation.

Kynell, Teresa C., and Bruce Seely. 2002. Historical Methods for Technical Communication. In *Research in Technical Communication*, ed. Laura J. Gurak and Mary M. Lay, 67–92. Westport: Praeger Publishers.

Longo, Bernadette. 2000. *Spurious Coin: A History of Science, Management, and Technical Writing*. New York: State University of New York Press.

Mather, Cotton, and Winton U. Solberg. 2000. *The Christian Philosopher*. Champaign: The University of Illinois Press.

Mokyr, Joel. 2006. The Great Synergy: The European Enlightenment as a Factor in Modern Economic Growth. In *Understanding the Dynamics of a Knowledge Economy*, ed. Wilfred Dolfsma and Luc Soete, 7–41. Cheltenham: Edward Elgar Publishing.

Moran, Michael G. 1985. The History of Technical and Scientific Writing. In *Research in Technical Communication: A Bibliographic Sourcebook*, ed. Michael G. Moran and Debra Journet, 25–38. Westport: Greenwood Press.

Moran, Michael G., and Debra Journet. 1985. Preface. In *Research in Technical Communication: A Bibliographic Sourcebook*, ed. Michael G. Moran and Debra Journet, i–xv. Westport: Greenwood Press.

Moran, Michael G., and Elizabeth Tebeaux. 2011. A Bibliography of Works Published in the History of Professional Communication from 1994–2009: Part 1. *Journal of Technical Writing and Communication* 41(2): 193–214.

Ong, Walter J. 2002. *Orality and Literacy: The Technologizing of the Word*. London: Routledge.

Pannabecker, John R. 1994. Diderot, the Mechanical Arts, and the Encyclopédie: In Search of the Heritage of Technology Education. *Journal of Technology Education* 6(1): 45–57.

Rickard, Thomas A. 1908. *A Guide to Technical Writing*. San Francisco: Mining and Scientific Press.

Rivers, William E. 1999. Studies in the History of Business and Technical Writing. In *Three Keys to the Past: The History of Technical Writing*, ed. Teresa C. Kynell and Michael G. Moran, 249–304. Stamford: Ablex Publishing Corporation.

Russell, Terence M. 1997. *The Encyclopaedic Dictionary in the Eighteenth Century*. Aldershot: Ashgate Publishing.

Savage, Gerald J. 2003. Towards Professional Status in Technical Communication. In *Power and Legitimacy in Technical Communication: The Historical and Contemporary Struggle for Professional Status*, ed. Teresa Kynell-Hunt and Gerald J. Savage, 1–13. New York: Baywood Publishing Company.

Stearns, Peter N. 2007. *The Industrial Revolution in World History*. 2nd ed. London: Westview Press.

Stevens, Edward W. 1995. *The Grammar of the Machine: Technical Literacy and Early Industrial Expansion in the United States*. New Haven: Yale University Press.

Tebeaux, Elizabeth. 1997. *The Emergence of a Tradition: Technical Writing in the English Renaissance, 1475–1640*. Amityville: Baywood Publishing Company.

———. 2008. Technical Writing in English Renaissance Shipwrightery: Breaching the Shoals of Orality. *Journal of Technical Writing and Communication* 38(1): 3–25.

———. 2014. *The Flowering of a Tradition: Technical Writing in England 1641–1700*. Amityville: Baywood Publishing Company.

Tebeaux, Elizabeth, and M. Jimmie Killingsworth. 1992. Expanding and Redirecting Historical Research in Technical Writing: In Search of Our Past. *Technical Communication Quarterly* 1(2): 5–32.

Thomas, Margaret, Gloria Jaffe, Peter Kincaid, and Yvette Stees. 1992. Learning to Use Simplified English: A Preliminary Study. *Technical Communication* 39(1): 69–73.

Von Tunzelmann, Nick. 1994. Technology in the Early Nineteenth Century. In *The Economic History of Britain since 1700, Vol. 1: 1700–1860*, 2nd ed., ed. Roderick Floud and Deirdre McCloskey, 271–299. Cambridge: Cambridge University Press.

Wasson, Ellis. 2009. *A History of Modern Britain: 1714 to the Present*. New York: John Wiley & Sons.

Weiss, Edmond H. 2005. *The Elements of International English Style*. New York: M.E. Sharpe.

White, Robert. 2001. *Discovering Old Cameras, 1839–1939*. 3rd ed. Botley: Osprey Publishing.

Wilson Sewing Machine Company. 1870. *Directions for Using the New Buckeye Under-Feed Sewing Machine*. Cleveland: Wilson Sewing Machine Company.

Woodward, Llewellyn. 1963. *The Age of Reform, 1815–70*. 2nd ed. Oxford: Oxford University Press.

Yeo, Richard. 2001. *Encyclopaedic Visions: Scientific Dictionaries and Enlightenment Culture*. Cambridge: Cambridge University Press.

2

Existing Controlled Languages for Technical Documents

Abstract The chapter first introduces the simplified international language Basic English. This was not developed specifically for technical documents. However, it is potentially adaptable to them. In addition, it provided an impetus and framework for the development of controlled languages specifically for technical documents. The chapter next chronologically introduces a number of controlled languages that have been used since the 1970s by large manufacturing companies, particularly to help make the information in their technical documents understandable. The chapter concludes by considering the benefits and drawbacks of controlled languages for technical documents.

Keywords Basic English • Bull Controlled English • Caterpillar Fundamental English • Ericsson English • Nortel Standard English • Perkins Approved Clear English

© The Author(s) 2017
S. Crabbe, *Controlling Language in Industry*,
DOI 10.1007/978-3-319-52745-1_2

Introduction

Historically, the notion of simplifying the international dissemination of technical knowledge can be traced as far back as the pre-industrial seventeenth century. Many simplified international languages were developed during the seventeenth century such as Universal Character (1657), Universal Language (1661), Ars Signorum (1661) and Philosophical Language (1668). However, what makes Logopandecteision (1653) different from the other simplified international languages being developed at this time is that its developer, Urquhart, states that it is "most fit for such as would with ease attaine to a most expedite facility of expressing themselves ... [concerning] ... mechanick trades" ([1653] 1834: 191). Urquhart describes Logopandecteision as having a simplified set of rules and controlled vocabulary "concerning mechanick trades in their tooles or tearmes" ([1653] 1834: 200). Unfortunately, Urquhart does not specify the rules or controlled vocabulary, so there is no way of knowing the applicability of Logopandecteision to the seventeenth-century mechanick trades. Nevertheless, it is clear from the previous chapter that the international dissemination of technical knowledge was still very limited when Urquhart developed Logopandecteision.

From the middle of the eighteenth century, the gradual adoption of British technology across continental Europe did provide an impetus for the emergence of a language for the international dissemination of technical knowledge. This was the English language. The British industrial revolution was then followed by the American industrial revolution which consolidated and expanded this use of English. By the late nineteenth and early twentieth century—commonly referred to by researchers as the period of the second industrial revolution (for example, Misa 2004; Muddiman 2008)—production in Britain, continental Europe and the USA was increasingly being internationalised and technologised. It was against this background that Ogden developed a simplified international language called Basic English which promised "an ideal of technological efficiency" (Russo 1989: 397) in the language sphere.

This chapter first introduces Basic English. It then chronologically introduces a number of controlled languages that have been used by large manufacturing companies since the 1970s for their technical documents. Specifically, it describes why and how each controlled language was developed and used. The chapter ends by looking at the benefits and drawbacks of controlled languages. It should be noted that there are controlled languages based on languages other than English. These include GIFAS Rationalised French, ScaniaSwedish and Siemens Dokumentationsdeutsch. This chapter focuses on English-based controlled languages while recognising that controlled languages based on other languages do exist.

Basic English

In the early twentieth century, the name Ford was synonymous with international uniformity and simplicity. With regard to cars, Ford declared "Any customer can have a car painted any colour that he wants so long as it is black" (1923: 72). With regard to language, Ford declared "Make everybody speak English" (cited by Ogden and Gordon 1994: 229).

One man who set out to achieve the latter was the British linguist Ogden, who in 1930 developed an international language called Basic English with a simplified set of rules and controlled vocabulary. Basic English had three main purposes: to serve as a simple common language for international communication, to allow non-English users to learn English in the shortest time possible and to regularise the English of native English users.

Ogden developed Basic English by reducing all the rules and vocabulary of English to just ten rules and 850 words. For expository and comparison purposes, the rules of Basic English, and the controlled languages introduced in this book, are classified into six broad categories according to their purpose. Fig. 2.1 explains and illustrates these categories. It should be noted from the outset that these categories are not intended to be definitive and some rules may be classified under more than one heading. Bloor and Bloor illustrate the difficulty of classifying language rules in their comment that "linguistic items, being multi-functional, can be looked at from more than one point of view, and hence given more than one label on different occasions even within the same analytical framework … so analysts may disagree on how to classify items without anyone necessarily being wrong or, for that matter, entirely right" (2004: 18).

Rule category	Rule purpose	Example rule
Grammatical	to influence how words are morphologically formed and how punctuation is used	Avoid conditional tenses.
Information Load	to influence the amount of information in sentences	Avoid too many subjects in one sentence.
Information Structure	to influence how the information content is structured	List events in sequence.
Lexical	to influence how words are selected and defined	Avoid abbreviations and colloquialisms.
Stylistic	to influence language style	Make positive statements.
Syntactic	to influence how sentences are structured	Use parallel construction.

Fig. 2.1 Categories of controlled language rules

The Basic English rules can be classified as grammatical (rules 1, 4, 5, 6 and 8), lexical (rules 2, 3, 7 and 10) and syntactic (rule 9). The rules, together with examples of how words can be formed using them, are as follows:

1. Add "s" to make the regular plural form. Apply standard rules for irregular plural forms such as "ies" and "es."
2. Make compounds by combining two nouns (for example, key + board → keyboard/book + mark → bookmark) or a noun and a directive particle (for example, on + line → online/ up + date → update).
3. Add "er" or "ing" to nouns to make the thing or person performing an operation or the operation itself (for example, work → worker and working/start → starter and starting).
4. Add "ing" or "ed" to nouns to make present-participle and past-participle adjectives (for example, work → working machine/measure → measured amount).
5. Add "more" or "most" to adjectives to express degree (for example, frequent → more frequent or most frequent). Learn the non-standard

adjectives that take "er" or "est" (for example, good → better and best).
6. Add "ly" to adjectives to make adverbs of manner (for example, bad → badly). Learn the conjugation of non-standard adjectives.
7. Add "un" to make negative adjectives (for example, necessary → unnecessary/important → unimportant).
8. Apply standard rules for conjugating verbs and pronouns.
9. Change the sentence word order and add "do" to make questions.
10. Use the English form for measurements, numbers, days, months and international words.

The first seven rules are concerned with lexical expansion through word formation techniques such as derivation and composition. Ogden developed Basic English as a multi-purpose simplified international language, as is reflected in the acronym *British American Scientific International Commercial*. The controlled vocabulary thus needed to have sufficient breadth and flexibility to enable its use for international communication in multiple domains. This was partly achieved through the first seven rules.

With regard to the controlled vocabulary, Ogden's research on word usage had identified a limited number of words as frequently occurring in dictionary definitions. This suggested the possibility of controlling vocabulary through restricting the number of words that can be used to this limited number of frequently occurring words. The 850 words Ogden identified became the Basic English controlled vocabulary. (For interested readers, these words can be viewed in full online at http://ogden.basic-english.org/words.html.)

Hinson describes Basic English as a "single meaning English lexicon and a simplified grammar" (1988: WE33) and Shubert et al. describe it as a "single meaning English lexicon along with simplified grammar rules" (1995: 348). However, Basic English is not a one-word-one-meaning language. Ogden's *Basic English: A General Introduction with Rules and Grammar* (1930) does not include any word definitions. Ogden did publish a dictionary of the controlled vocabulary titled *The Basic Words: A Detailed Account of Their Uses* (1932). However, its purpose was "to give an idea of the way in which the 850 words may be used" (Ogden 1932: V). The dictionary entries are thus intended only to be illustrative. In fact, turning to Fries and Traver (1950), one finds that the 850 words have 18,416 different meanings in the Oxford English Dictionary.

Davies suggests that "the purpose of Basic English was to promote mainly technical communication" (1999: 110). Sonntag similarly suggests that Basic English was "meant to serve as an international medium of communication, particularly in technology" (2003: 23). However, Basic English was, as earlier noted, developed for international communication in multiple domains. In fact, Ogden recognised that the Basic English controlled vocabulary was insufficient for the international dissemination of technical knowledge. Basic English thus allows its users to, first, extend the controlled vocabulary by a further 150 specialised technical terms and to, second, use "technical terms ... [that] ... are introduced into the text with explanatory matter" (Ogden 1940: 74). Nevertheless, the author has not yet found any evidence that manufacturing companies have used Basic English for their technical documents.

Interest in Basic English was considerable during the 1930s with *Basic English: A General Introduction with Rules and Grammar* going through eight editions between 1930 and 1940. Churchill was also a prominent supporter of Basic English, and the National Archives reveal that its possible use was discussed at a Cabinet meeting held in 1943. (For interested readers, the record of the Cabinet meeting can be viewed online at http://www.nationalarchives.gov.uk/releases/2006/january/january1/english.htm.) However, its popularity started to diminish during the Second World War and never fully recovered thereafter. It did, nonetheless, provide an impetus and framework for the development of Caterpillar Fundamental English (CFE).

Caterpillar Fundamental English

Caterpillar Fundamental English (CFE) is widely agreed to be the first English-based controlled language used by a manufacturing company for its technical documents (for example, Wojcik and Hoard 1997; van der Eijk 1998; Kaji 1999; Hartley and Paris 2001).

2 Existing Controlled Languages for Technical Documents

Caterpillar Inc. (henceforth Caterpillar) is an American manufacturer of engines and machines. Its successful expansion into international markets between 1950 and 1970 meant that, by 1971, it had 20,000 English language technical documents for maintaining and repairing its engines and machines and 10,000 service staff with 50 different native languages.

Caterpillar developed CFE so that the native English users, non-native English users and non-English users among its international service staff could all understand the English language maintenance and repair documents, thus eliminating the need to translate them into other languages. This kind of controlled language is classified by researchers such as Nyberg et al. (2003), O'Brien (2003) and Reuther (2003) as human-oriented in that its aim is to help make the information in technical documents understandable for human readers.

Caterpillar's von Glasenapp describes the manufacturing company's problem, and the impetus for developing CFE, as follows:

> To translate *all* publications into *all* languages is impossible … 20,000 pieces of literature times 50 languages would give 1 million different versions … [which] … is beyond current printing capabilities and any economical estimate … It is apparent that translation is no longer a feasible answer. (1972: 81)

Turning to van der Eijk (1998), one finds that Caterpillar initially considered using Basic English. However, it was rejected in favour of a new controlled language designed specifically for Caterpillar's maintenance and repair documents. Nevertheless, Basic English and CFE are similar in that Basic English has ten rules and an 850-word controlled vocabulary and CFE has ten rules and an 800-word controlled vocabulary. The ten CFE rules are as follows:

1. Make positive statements.
2. Avoid long and complicated sentences.
3. Avoid too many subjects in one sentence.
4. Avoid too many successive adjectives and nouns.
5. Use uniform sentence structures.
6. Avoid complicated past and future tenses.

7. Avoid conditional tenses.
8. Avoid abbreviations, contractions and colloquialisms.
9. Use punctuation correctly.
10. Use consistent nomenclature.

The rules can be classified as lexical (rules 8 and 10), syntactic (rules 2, 4 and 5), grammatical (rules 6, 7 and 9), stylistic (rule 1) and as addressing information load (rule 3). This breakdown is similar to that for the Basic English rules. However, the content of the individual rules is different, reflecting their different purpose. The rules control linguistic and organisational features of Caterpillar's maintenance and repair documents in order to help make the information in them understandable for the native English users, non-native English users and non-English users among Caterpillar's international service staff. They were developed specifically for Caterpillar's technical documents, but they do seem to be sufficiently general to have much broader utility. This suggests the potential for different manufacturing companies to have common or similar rules in their controlled languages.

The 800-word CFE controlled vocabulary was developed in a similar way to the 850-word Basic English controlled vocabulary in that it was derived from the most frequently recurring words in a broad sample of Caterpillar's maintenance and repair documents. In addition, as with Basic English, the controlled vocabulary can be extended by user-approved specialised technical terms. There is, nonetheless, a fundamental difference between CFE and Basic English in that each word in the CFE controlled vocabulary has an approved spelling, part(s) of speech and meaning(s) and cannot be used with any other spelling, grammatical form or sense. For example, drop is approved as a noun that means a small quantity of liquid, decrease as a verb that means to fall, right as an adjective that means the opposite of left and correct as an adjective that means the opposite of wrong. This was done to ensure that words were spelled and used uniformly both in and across Caterpillar's technical documents. Caterpillar determined these approved spellings, parts of speech and meanings from the most frequently recurring spellings, parts of speech and meanings in the broad sample of its maintenance and repair documents.

Caterpillar produced tens of thousands of pages of maintenance and repair documents using CFE between 1971 and 1982, and it is widely agreed that the information in these technical documents was understandable for the non-native English users and non-English users among Caterpillar's international service staff after English language courses of between just 30 and 60 hours (for example, Hinson 1988; Dekker and Wijma 2004; Kirkman 2005). In addition, von Glasenapp points out that the native English users among Caterpillar's international service staff benefitted from having the maintenance and repair documents at "a more universally understandable level" (1972: 84).

Caterpillar nevertheless stopped using CFE in 1982 because, first, the high turnover of service staff made running its English language courses expensive and time-consuming and, second, it began trailing a machine translation system developed by Carnegie Mellon University to translate some of its maintenance and repair documents into French, German and Spanish. This encouraged Caterpillar to develop a new controlled language with the dual aim of helping make the information in its maintenance and repair documents understandable and aiding its machine translation. This new controlled language was named Caterpillar Technical English (CTE) and is, at the time of writing, still in use.

The year that CFE was developed, von Glasenapp argued that "the theory of Caterpillar Fundamental English is widely applicable, and all indications are that this language system might soon spread" (1972: 84). Von Glasenapp's words were to prove prophetic with many manufacturing companies following Caterpillar and starting to use controlled languages for their own technical documents.

Perkins Approved Clear English

Perkins Engines (henceforth Perkins) is an American (formerly British) manufacturer of engines and machines. By the late 1970s, Perkins' successful expansion into international markets meant that it had service staff and customers in approximately 160 countries. Perkins Technical Publications Department in Britain produced all the English language installation documents for its engines and machines of which only some were translated into the languages of its international service staff and customers.

Perkins Approved Clear English (PACE) was developed by the Technical Publications Department in 1980. Pym, the manager at this time, describes the reason for its development:

> Our overall objective is to improve comprehension ... particularly for those persons using English instead of their native language, and to aid translation ... Technical Publications had been concerned about the quality of written communications for some time ... their content was often ambiguous and full of jargon. (1990: 80–82)

PACE was thus developed with the dual aim of helping make the information in Perkins' technical documents understandable and aiding its translation. In fact, five years later in 1985, Perkins successfully trialed the Weidner MicroCAT machine translation system to translate some of its PACE-compliant English language installation documents. Pym makes clear that the introduction of a machine translation system was "a logical consequence of using controlled English—which is its ideal input" (cited by Brockmann 1990: 113). This kind of controlled language is classified by researchers such as Nyberg et al. (2003), O'Brien (2003) and Reuther (2003) as dual-oriented in that its dual aim is to help make the information in technical documents understandable for human readers and translatable, particularly with machine translation systems.

PACE, like Basic English and CFE, has a simplified set of rules and controlled vocabulary. In the case of PACE, the same rules are used to help make the information in Perkins' installation documents understandable and aid its translation using the Weidner MicroCAT machine translation system. The ten PACE rules are as follows:

1. Keep sentences short.
2. Omit redundant words.
3. Order the parts of the sentence logically.
4. Don't change constructions in mid-sentence.
5. Take care with the logic of *and* and *or*.
6. Avoid elliptical constructions.

7. Don't omit conjunctions or relatives.
8. Adhere to the PACE dictionary.
9. Avoid strings of nouns.
10. Do not use -ing unless the word appears thus in the PACE dictionary.

The rules can be classified as lexical (rules 5 and 8), syntactic (rules 1, 3, 4, 6 and 9), grammatical (rules 7 and 10) and stylistic (rule 2). Most of the rules, as with most of the CFE rules, are thus lexical, syntactic and grammatical. In fact, there are similarities between the PACE and CFE rules with regard to the specific linguistic features of the technical documents that they control. By way of illustration, PACE states to "Keep sentences short." (rule 1) and CFE to "Avoid long and complicated sentences." (rule 2) and PACE states to "Avoid strings of nouns." (rule 9) and CFE to "Avoid too many successive adjectives and nouns." (rule 4).

Perkins developed an in-house dictionary of the controlled vocabulary that was regularly updated and expanded until March 2001. The March 2001 edition of this dictionary has 2750 entries if each part of speech is counted as one entry. Each entry has an approved spelling, approved part of speech and approved meaning (or limited number of meanings) as with CFE. For example, drop is approved as a noun that means "a quantity of fluid that falls in one spherical mass", decrease as a verb that means "to reduce", right as an adjective that means "when facing north, right is east" and correct as a verb that means "to remove fault". Perkins, as with Caterpillar, determined these approved spellings, parts of speech and meanings from the most frequently recurring spellings, parts of speech and meanings in a broad sample of its installation documents.

The PACE controlled vocabulary is notably larger than the Basic English and CFE controlled vocabularies. This is because, unlike Basic English and CFE, the PACE controlled vocabulary cannot be extended by user-approved specialised technical terms. All approved terms are included in the dictionary.

Before going further, it might be useful for readers to look at an example of a controlled language in use. The following shows an extract from a Perkins' installation document before and after compliance with PACE:

Reference to renewing joints and cleaning of joint faces has to a great extent been omitted from the text, it being understood that this will be carried out where applicable.
Normally the text does not include instructions to clean joint faces or to renew joints. Where they are relevant, these operations must be done. (Pym 1990: 88)

The original 30-word sentence has been broken up into two short sentences in compliance with the rule "Keep sentences short." (rule 1). The redundant words "it being understood that" have been removed in compliance with the rule "Omit redundant words." (rule 2). The phrase "renewing joints and cleaning of joint faces" has been rewritten as "to clean joint faces or to renew joints" to make clear that it refers to two distinct operations rather than one operation and one type of joint called a renewing joint. This is in compliance with the rule "Take care with the logic of *and* and *or*." (rule 5). Finally, the verb "carried out" is not in the PACE dictionary. It has been replaced with "done" in compliance with the rule "Adhere to the PACE dictionary." (rule 8) and the PACE dictionary entry for done: "(vb) To bring to pass".

Pym (1990) reveals that using PACE helped make the information in Perkins' installation documents understandable and aided its translation. However, Caterpillar purchased Perkins in 1997 and the Technical Publications Department changed to using Caterpillar Technical English (CTE) six years later.

The nature of the discussion thus far might leave readers with the impression that controlled languages are only used by manufacturing companies headquartered in countries with English as the native and/or official language. However, this is not the case.

Ericsson English

Telefonaktiebolaget LM Ericsson (henceforth Ericsson) is a Swedish manufacturer of industrial telecommunications equipment and machines. In the late 1970s, Ericsson approached the Communication Studies Unit of the University of Wales Institute of Science and Technology (now Cardiff

2 Existing Controlled Languages for Technical Documents

University) to develop a controlled language for its English language service and maintenance documents. The result was Ericsson English (EE). The nature of Ericsson's problem at this time is described in the introduction to the *Ericsson English Writer's Guide* (1983):

> Ericsson English (EE) was created for use in basic technical instructions and descriptions. It was created to help make those basic documents comprehensive to readers who have a poor command of English. (Telefonaktiebolaget LM Ericsson 1983: 3)

EE—like CFE and PACE—has a simplified set of rules and controlled vocabulary. In the case of EE, there are 23 rules. This is more than CFE and PACE. The rules are thus often more specific and detailed. The 23 EE rules are as follows:

1. Do not use synonyms.
2. Do not make long strings of nouns or adjectives before a noun.
3. Do not add prefixes or suffixes to EE words to make new words.
4. Do not use conversational expressions (idioms, contractions, slang).
5. Do not use unnecessary abbreviations.
6. Do not add new words to the Word List unnecessarily.
7. Write clearly and simply.
8. Use only the active voice of verbs.
9. Use only the simple past, simple present and simple future tenses. Always use the present tense when possible.
10. EE has only four auxiliary verbs: "must," "can," "will" and "do."
11. Avoid unnecessary negatives.
12. Form questions as in ordinary English.
13. Try not to use too many adverbial clauses in one sentence. If you start a sentence with an adverbial clause, use a comma to mark the end of the clause.
14. Use "which" only as an interrogative adjective. You cannot use "which" to introduce an adverbial clause. You must use "that" instead.
15. Write short sentences dealing with just one idea.
16. "-ing" words are nouns that describe activities.

"-ed" words are adjectives. Do not use them to form the passive voice or the perfect tense.
17. Use correct and consistent punctuation.
18. Always use a comma between a condition and an instruction or statement.
19. Use a colon between two parts of a sentence when the second part is an example, list or explanation of the first part.
20. Use a hyphen between two words if it makes the meaning clearer.
21. Use dashes to mark a list of ideas after a colon. Do not use dashes in other ways.
22. Use an exclamation mark to stress the importance of a warning.
23. Use an oblique mark (/) to mean and/or, not from/to.

The rules can, however, still be similarly classified as lexical (rules 1, 2, 3, 4, 5, 6, 14 and 20), syntactic (rules 12, 13, 19, 21 and 23), grammatical (rules 8, 9, 10, 16, 17 and 18), stylistic (rules 7, 11 and 22) and as addressing information load (rule 15). They control linguistic and organisational features of Ericsson's service and maintenance documents with the aim of making the information in them understandable for those of Ericsson's staff with limited English language skills.

The *Ericsson English Writer's Guide* (1983) includes a dictionary of the controlled vocabulary. This has 2100 entries if each part of speech is counted as one entry. Each entry has an approved spelling, approved part of speech and approved meaning (or limited number of meanings) as with CFE and PACE. For example, drop is approved as a noun that means "a quantity of fluid that falls in one spherical mass", decrease as a verb that means "to reduce, to fall in value", right as an adjective and noun that means "when facing north, right is east" and correct as a verb that means "to take out mistakes from, to restore to proper state". However, in a departure from CFE and PACE, this dictionary was not developed from scratch. Rather, the University of Wales Institute of Science and Technology purchased a commercially available controlled language called International Language for Service and Maintenance (ILSAM) from M. and E. White Consultants and modified its controlled vocabulary to create the controlled vocabulary in the *Ericsson English Writer's Guide* (Telefonaktiebolaget LM Ericsson 1983: introduction). It is, however, interesting to note that researchers agree that ILSAM was itself developed from CFE (for example, Kirkman et al. 1980; Schreurs and Adriaens 1992; Newton 1992; Shubert et al. 1995; Kirkman 2005).

The *Ericsson English Writer's Guide* cautions technical communicators that they "must conform to the rules strictly ... otherwise the claims we make for the readability of EE will not be justifiable" (Telefonaktiebolaget LM Ericsson 1983: introduction). One of these claims is that "Ericsson English is so easy ... that other languages should not be necessary" (Telefonaktiebolaget LM Ericsson 1983: introduction).

Ericsson thus departed from Caterpillar and Perkins in not developing its controlled vocabulary from scratch. However, it has not been the only manufacturing company to have done this.

Nortel Standard English

Nortel Networks Corporation (henceforth Nortel) is a Canadian manufacturer of industrial telecommunications equipment and machines. In 1993, Calistro from Nortel's Global Publishing Technology unit stated that "The increase in the number of products with accompanying documentation sold around the world has shown the need to develop some form of controlled English" (1993: 158). The following year Calistro approached Smart Communications Inc. (henceforth Smart Communications) to develop a controlled language to make the information in Nortel's technical documents understandable and aid its translation. The result was Nortel Standard English (NSE).

NSE—like CFE, PACE and EE—has a simplified set of rules and controlled vocabulary. In the case of NSE, there are 15 rules, which are as follows:

1. **Write positive statements.** Use positive statements instead of negative statements when possible. Exception: Use negative statements in warnings and cautions.
2. **Write short sentences.** Write sentences that have less than 22 words.

3. **Write simple sentences that have one thought per sentence.** Write sentences with simple sentence structure. Try to have one thought per sentence.
4. **List events in sequence.** Write events in the order in which they occur.
5. **Use parallel structure.** Use parallel grammatical structure for a series of items in both sentences and bulleted lists.
6. **Use the active voice.** Use the active voice instead of the passive voice.
7. **Do not use a gerund at the beginning of a sentence or in an ambiguous construction.**
8. **Do not use could, should, would, might and may.** Avoid the ambiguous helping verbs could, should, would, might and may.
9. **Use the present tense instead of the future tense.** Exception: Use the future tense if the action is a true future action.
10. **Avoid noun strings of more than three words.** Do not use noun strings that have more than three words.
11. **Use only NSE approved abbreviations and acronyms.** Use only abbreviations and acronyms that are in the NSE dictionaries. If you need to create an abbreviation or acronym, see your NSE committee representative.
12. **Use specific terms or words.** Make sure that every pronoun has a clear reference. Do not use this or that as a pronoun.
13. **Do not use jargon and idiomatic expressions.** Use only terms that are in the NSE dictionaries.
14. **Use appropriate language.** Avoid the following types of words:

    a. Gender-specific, racial or religious words
    b. Words specific to a region or dialect
    c. Subjective, emotive or negative words
    d. Words that are not clear

15. **Use a determiner with nouns and noun phrases when possible.**

The rules can be classified as lexical (rules 10, 11, 12, 13 and 14), syntactic (rules 2 and 5), grammatical (rules 6, 7, 8, 9 and 15), stylistic (rule 1) and as addressing information load (rule 3) and information structure (rule 4). The third rule can, in fact, be broken down into two sub-rules of which the first "Write sentences with simple sentence structure." is syntactic and the second "Try to have one thought per sentence." addresses information load. The rules control linguistic and organisational features of Nortel's technical documents to make the information in them understandable and aid its translation. Most of these rules are lexical, syntactic and grammatical as with most of the CFE, PACE and EE rules.

The NSE controlled vocabulary is identical to the CFE, PACE and EE controlled vocabularies in that words have an approved spelling, approved part of speech (or limited number of parts of speech) and approved meaning (or limited number of meanings). As with Ericsson, however, Nortel did not develop the NSE controlled vocabulary from scratch. Rather, Nortel purchased the commercially available controlled language checker MAXit from Smart Communications and created NSE's controlled vocabulary from the customisable MAXit dictionaries.

According to Smart from Smart Communications, the use of NSE and MAXit "made complex telecommunications documentation easier to read and understand" (2006: 3). However, Nortel declared bankruptcy in 2009 and NSE is no longer used.

It was suggested earlier in this chapter that different manufacturing companies may share common or similar controlled language rules. This is explored in more detail in the next section.

Bull Controlled English

The final controlled language introduced in this chapter is Bull Controlled English (BCE). Groupe Bull (henceforth Bull) is a French manufacturer of computers and consumer electronics. In the early 1990s, Bull decided to develop the technical documents for the setup, operation and maintenance of its key goods first in English and then in other languages in order to reduce the launch time of these goods to international markets. This decision was, in the words of Joscelyne, "an implicit recognition of the pre-eminence of English in the computing sector" (1992: 4).

At the same time, Bull's ILO group (l'Internationalisation et Localisation de l'Offre Bull) introduced BCE for its English language technical documents and the SYSTRAN machine translation system to translate them into Dutch, French, German, Italian and Spanish. The primary aim of BCE was to help make the information in Bull's English language technical documents understandable for, first, native English users and, second, non-native English users with languages other than Dutch, French, German, Italian and Spanish. The nature of Bull's problem is described by Lee, a Bull manager at this time:

> In an international market place, many users of documentation have English as their second, or even third, language. Technical documentation should therefore use simple grammatical structures ... [Additionally] ... particularly in the field of computers, technology has become available to a wider public. The end user is not necessarily a ... highly literate person and requires a simplified English. (1993: 35–36)

BCE, like the other controlled languages introduced in this chapter, has a simplified set of rules and controlled vocabulary. The ten BCE rules are as follows:

1. Make positive statements: avoid the passive voice; avoid the future tense.
2. Keep sentence length to a maximum of 25 words.
3. Use valid technology; do not invent it. Use the Controlled English vocabulary.
4. One thought per sentence.
5. Use simple sentence structures.
6. Use parallel construction.
7. Avoid conditional tenses.
8. Avoid abbreviations and colloquialisms.
9. Use correct punctuation.
10. Use the tools available (Max. Grammar Checker, Spelling Checker).

The first rule "Make positive statements: avoid the passive voice; avoid the future tense." can be broken down into three sub-rules of which the first is stylistic and the second and third are grammatical. The second to ninth rules can be classified as lexical (rules 4 and 8), syntactic (rules 2, 3 and 6), grammatical (rules 7 and 9) and as addressing information load (rule 5). These nine rules control linguistic and organisational features of Bull's technical documents to help make the information in them under-

2 Existing Controlled Languages for Technical Documents

standable. The final rule "Use the tools available (Max. Grammar Checker, Spelling Checker)." refers to software tools that were developed to support Bull's technical communicators. Most of the rules, as with most of the rules from the other controlled languages introduced in this chapter, are lexical, syntactic and grammatical. In addition, their content seems similar to the other controlled languages introduced in this chapter. In order to check this, Fig. 2.2 provides a comparison of the BCE and CFE rules.

BCE rules		CFE rules
Make positive statements: avoid the passive voice; avoid the future tense. (rule 1)	⇔	Make positive statements. (rule 1)
Keep sentence length to a maximum of 25 words. (rule 2)	⇔	Avoid long and complicated sentences. (rule 2)
Use simple sentence structures. (rule 5)		
Use valid terminology: do not invent it. Use the Controlled English vocabulary. (rule 3)	⇔	Use consistent nomenclature. (rule 10)
One thought per sentence. (rule 4)	⇔	Avoid too many subjects in one sentence. (rule 3)
Use parallel construction. (rule 6)	⇔	Use uniform sentence structures. (rule 5)
Avoid conditional tenses. (rule 7)	⇔	Avoid conditional tenses. (rule 7)
Avoid abbreviations and colloquialisms. (rule 8)	⇔	Avoid abbreviations, contractions, and colloquialisms. (rule 8)
Use correct punctuation. (rule 9)	⇔	Use punctuation correctly. (rule 9)
Use the tools available (Max. Grammar Checker, Spelling Checker). (rule 10)		
		Avoid complicated past and future tenses. (rule 6)
		Avoid too many successive adjectives and nouns. (rule 4)

Fig. 2.2 Comparison of the BCE and CFE rules

This reveals that there is a very high degree of similarity between the BCE and CFE rules with only one unique BCE rule and two unique CFE rules. Furthermore, the one unique BCE rule "Use the tools available (Max. Grammar Checker, Spelling Checker)." (rule 10) is not applicable to CFE as there were no software tools developed for it.

The Bull technical communicators were provided with three dictionaries: a general dictionary which provided the controlled vocabulary, a technical dictionary which provided additional Bull-approved technical terms and a synonym dictionary which provided Bull-approved synonyms from a list of commonly used synonyms in English language technical documents for computers and consumer electronics. Spelling, part of speech and meaning were all controlled as with the other controlled languages. In the case of BCE, the general and synonym dictionaries were purchased from Smart Communications, which, readers will recall, is the same organisation that Nortel purchased the commercially available controlled language checker MAXit from.

Bull was partially privatised in 1996 with ownership divided between the French government and companies from France (France Telecom), Japan (NEC and Dai Nippon Printing) and the USA (Motorola). As a result of the privatisation, BSE stopped being used.

Benefits and Drawbacks of Controlled Languages

There is a widespread and strong agreement in the literature that the use of controlled languages can help make the information in technical documents understandable and aid its machine translation.

In terms of understandability, it is agreed that they help make the information in technical documents understandable for both native and non-native English users (for example, Brockmann 1990; Arnold et al. 1994; Brockmann 1997; Reuther 1998; Haller and Schütz 2001; Markantonatou et al. 2002; Nyberg et al. 2003; Rychtyckyj 2006). However, it should be pointed out that some of this agreement is based on generalisations such as "it goes without saying that controlled language makes it easier ... to understand a text" (Brockmann 1997: 10) or "it is a well known and indisputable fact within the CL community that the use of a Controlled Language (CL) in technical documentation leads to quality improvement with respect to readability" (Reuther 1998: 174).

In terms of translatability, it is agreed that they aid translation, particularly with machine translation systems (for example, Lee 1993; Arnold et al. 1994; Murphy et al. 1998; Kaji 1999; Esselink 2000; Weiss 2005).

It should also be noted in conclusion that there is some argument in the literature that it can be difficult for technical communicators to use controlled languages because of the restrictions imposed (for example, Hoard et al. 1992; Thomas et al. 1992; Douglas and Hurst 1996). However, this argument is not universally accepted, with Joshi (2006) and Smart (2006) counter-arguing that it is not difficult to use controlled languages.

Summary

The main objective of this chapter has been to chronologically introduce controlled languages that have been used by large manufacturing companies for their English language technical documents since the 1970s. Specifically, it has looked at why and how each controlled language was developed and used. In doing this, it will have become clear to readers that controlled languages restrict certain linguistic and organisational features of technical documents to help make the information in them understandable and, in many cases, simultaneously aid its machine translation.

While controlled languages have been developed for manufacturing companies, knowledge of the specific linguistic and organisational features of modern technical documents that help make the information in them understandable has been greatly enhanced by the publication of an extensive range of government, academic and professional literature. A complete understanding of controlled languages thus necessitates an understanding of these linguistic and organisational features of technical documents, which are explored in the next chapter of this book.

Bibliography

Arnold, Doug J., Lorna Balkan, R. Lee Humphreys, Siety Meijer, and Louise Sadler. 1994. *Machine Translation: An Introductory Guide*. Oxford: NCC Blackwell.

Bloor, Thomas, and Meriel Bloor. 2004. *The Functional Analysis of English*. 2nd ed. London: Arnold Publishers.

Brockmann, Daniel. 1997. Controlled Language and Translation Memory Technology: A Perfect Match to Save Translation Cost. *TC-Forum* 4(97): 10–11.

Brockmann, R. John. 1990. *Writing Better Computer User Documentation: From Paper to Hypertext*. New York: John Wiley & Sons, Inc..

Calistro, Ralph. 1993. Simplified English Roundup: Fait Accompli or Impossible Dream? In *Proceedings of the 40th International Technical Communication Conference, Dallas, Texas*, ed. Society for Technical Communication, 158–160. Washington, DC: Society for Technical Communication.

Davies, Alan. 1999. *An Introduction to Applied Linguistics: From Theory to Practice*. Edinburgh: Edinburgh University Press.

Dekker, Jan M., and Frans Wijma. 2004. *Simplified English: The New Language in International Business*. 2nd ed. Tilburg: Tedopres International B.V.

Douglas, Shona, and Matthew Hurst. 1996. Controlled Language Support for Perkins Approved Clear English (PACE). In *Proceedings of the First International Workshop on Controlled Language Applications, CLAW 96*, 93–105. Leuven: Katholieke Universiteit Leuven.

Esselink, Bert. 2000. *A Practical Guide to Localization*. Amsterdam: John Benjamins Publishing Company.

Ford, Henry. 1923. *My Life and Work*. Garden City: Doubleday, Page & Company.

Fries, Charles C., and Alice A. Traver. 1950. *English Word Lists: A Study of Their Adaptability for Instruction*. Ann Arbor: George Wahr Publishing Company.

Haller, Johann, and Jörg Schütz. 2001. CLAT: Controlled Language Authoring Technology. In *Proceedings of the 19th Annual International Conference on Computer Documentation*, ed. Mary J. Northrop and Scott Tilley, 78–82. New York: ACM Press.

Hartley, Anthony, and Cécile Paris. 2001. Translation, Controlled Languages, Generation. In *Exploring Translation and Multilingual Text Production: Beyond Context*, ed. Erich Steiner and Colin Yallop, 307–326. Berlin: Walter de Gruyter.

Hinson, Don E. 1988. Simplified English—Is It Really Simple? In *Proceedings of the 38th International Technical Communication Conference*, ed. Society for Technical Communication, WE33–WE36. Washington, DC: Society for Technical Communication.

Hoard, James, Richard Wojcik, and Katherina Holzhauser. 1992. An Automated Grammar and Style Checker for Writers of Simplified English. In *Computers and Writing: State of the Art*, ed. Patrick Holt and Noel Williams, 278–296. Oxford: Intellect Books.

Joscelyne, Andrew. 1992. Localizing the Bull Way. *Language Industry Monitor* 9: 4–5.

Joshi, Rajmohan. 2006. *Writing Skills for Technical Purposes*. Delhi: Isha Books.

Kaji, Hiroyuki. 1999. Controlled Languages for Machine Translation: State of the Art. In *Proceedings of the Machine Translation Summit VII*, ed. Miriam Butt and Tracy H. King, 37–39. Singapore: The Asia-Pacific Association for Machine Translation.

Kirkman, John. 2005. *Good Style: Writing for Science and Technology*. 2nd ed. London: Routledge.

Kirkman, John, Christine Snow, and Ian Watson. 1980. Controlled English in International Documentation. In *New Trends in Documentation and Information: Proceedings of the 39th FID Congress*, ed. Peter J. Taylor, 70–76. London: Aslib.

Lee, Arthur. 1993. Controlled English with and Without Machine Translation. In *Translating and the Computer 15*, ed. The Association for Information Management, 35–39. London: Aslib.

Markantonatou, Stella, Vangelis Karkaletsis, and Yanis Maistros. 2002. An Authoring Tool for Controlled Modern Greek. In *Proceedings of the 2nd Hellenic Conference on Artificial Intelligence (SETN-2002)* (Companion Volume), ed. Ioannis Vlahavas and Constantine Spyropoulos, 165–176. Thessaloniki: Aristotle University of Thessaloniki.

Misa, Thomas J. 2004. *Leonardo to the Internet: Technology and Culture from the Renaissance to the Present*. Baltimore: John Hopkins University Press.

Muddiman, Dave. 2008. Public Science in Britain and the Origins of Documentation and Information Science, 1890–1950. In *European Modernism and the Information Society: Informing the Past, Understanding the Present*, ed. Warden B. Rayward, 201–222. Aldershot: Ashgate Publishing.

Murphy, Dawn, Jane Mason, and Stuart Sklair. 1998. Improving Translation at the Source. In *Translating and the Computer 20*, ed. The Association for Information Management, 1–8. London: Aslib.

Newton, John. 1992. The Perkins Experience. In *Computers in Translation: A Practical Appraisal*, ed. John Newton, 46–57. London: Routledge.

Nyberg, Eric H., Teruko Mitamura, and Willem-Olaf Huijsen. 2003. Controlled Language for Authoring and Translation. In *Computers and Translation*, ed. Harold Somers, 245–282. Amsterdam: John Benjamins Publishing Company.

O'Brien, Sharon. 2003. Controlling Controlled English: An Analysis of Several Controlled Language Rule Sets. In *Proceedings of EAMT-CLAW 2003*, 105–114. Dublin: Dublin City University.

Ogden, Charles K. 1932. *The Basic Words: A Detailed Account of Their Uses*. London: Kegan Paul, Trench, Trubner & Company.

———. 1940. *Basic English: A General Introduction with Rules and Grammar*. 8th ed. London: Kegan Paul, Trench, Trubner & Company.

Ogden, Charles K., and Terrence W. Gordon. 1994. *C.K. Ogden and Linguistics*. London: Routledge.

Pym, Peter J. 1990. Pre-Editing and the Use of Simplified Writing for MT: An Engineer's Experience of Operating an MT System. In *Translating and the Computer 10*, ed. The Association for Information Management, 80–96. London: Aslib.

Reuther, Ursula. 1998. Controlling Language in an Industrial Application. In *Proceedings of the Second International Workshop on Controlled Language Applications, CLAW 98*, 174–184. Pittsburgh: Language Technologies Institute, Carnegie Mellon University.

———. 2003. Two in One-Can It Work? Readability and Translatability by Means of Controlled Language. In *Proceedings of EAMT/CLAW 2003*, 124–132. Dublin: EAMT/CLAW.

Russo, John P. 1989. *I.A. Richards: His Life and Works*. London: Routledge.

Rychtyckyj, Nestor. 2006. Standard Language at Ford Motors: A Case Study in Controlled Language Development and Deployment. *Machine Translation Archive*. http://www.mt-archive.info/CLAW-2006-Rychtyckyj.pdf. Accessed 5 September 2016.

Schreurs, Dirk, and Geert Adriaens. 1992. Controlled English (CE): From COGRAM to ALCOGRAM. In *Computers and Writing: State of the Art*, ed. Patrik Holt and Noel Williams, 206–221. Oxford: Intellect Books.

Shubert, Serena, Heather Holmback, Jan Spyridakis, and Mary Coney. 1995. The Comprehensibility of Simplified English in Procedures. *Journal of Technical Writing and Communication* 25(4): 347–369. Machine Translation Archive

Smart, John M. 2006. Smart Controlled English—Paper and Demonstration. *Machine Translation Archive.* http://www.mt-archive.info/CLAW-2006-Smart.pdf. Accessed 8 September 2016.

Sonntag, Selma K. 2003. *The Local Politics of Global English: Case Studies in Linguistic Globalization.* Lanham: Lexington Books.

Telefonaktiebolaget LM Ericsson. 1983. *Ericsson English Writer's Guide.* Stockholm: Telefonaktiebolaget LM Ericsson Standards Department.

Thomas, Margaret, Gloria Jaffe, Peter Kincaid, and Yvette Stees. 1992. Learning to Use Simplified English: A Preliminary Study. *Technical Communication* 39(1): 69–73.

Urquhart, Thomas. [1653] 1834. *The Works of Sir Thomas Urquhart, of Cromarty, Knight.* Reprinted from the Original Editions. Edinburgh: The Maitland Club.

Van der Eijk, Pim. 1998. Controlled Languages in Technical Documentation. In *Computational Linguistics in the Netherlands 1997: Selected Papers from the Eighth CLIN Meeting,* ed. Peter-Arno Coppen, Hans van Halteren, and Lisanne Teunissen, 187–204. Amsterdam: Rodopi.

Von Glasenapp, Bernt. 1972. Caterpillar Fundamental English. In *Proceedings of the 19th International Technical Communication Conference,* ed. Society for Technical Communication, 81–85. Arlington: Society for Technical Communication.

Weiss, Edmond H. 2005. *The Elements of International English Style.* New York: M.E. Sharpe.

Wojcik, Richard H., and James E. Hoard. 1997. Controlled Languages in Industry. In *Survey of the State of the Art in Human Language Technology,* ed. Ron Cole, Joseph Mariani, Hans Uszkoreit, Giovanni B. Varile, Annie Zaenen, Antonio Zampolli, and Victor Zue, 238–243. Cambridge: Cambridge University Press.

3

Best-Practice Features of Modern Technical Documents

Abstract The chapter first describes the most widely agreed upon linguistic and organisational best-practice features of modern technical documents that help make the information in them understandable. These are identified from academic, government and professional literature published over the last two and a half decades. The chapter concludes by identifying the key benefits of developing understandable technical documents.

Keywords Modern technical documents • Technical document features • Linguistic features • Organisational features • Understandable technical documents

Introduction

> … if you go and look at the instruction manual you would have thought that somehow there would be a correspondence between the amazingness of the piece of technology and the amazingness of the instruction manual, but it's not true. (Martin [Producer], How to Write an Instruction Manual, 2009)

© The Author(s) 2017
S. Crabbe, *Controlling Language in Industry*,
DOI 10.1007/978-3-319-52745-1_3

On a BBC Radio 4 programme titled *How to Write an Instruction Manual* that was broadcast on August 21, 2009, Miodownik criticised the quality of the English language technical documents provided by manufacturing companies for their consumer technology. This kind of criticism of technical documents is not confined to Miodownik. Over the last two and a half decades alone, Brockmann (1990), Cohen (1995), Schriver (1997), Council of the European Union (1998), Thimbleby (2000), Griffiths (2001), Warren (2001), Hargis et al. (2004), Byrne (2006), Cutts (2009), Byrne (2012) and Whitaker and Mancini (2013) have all made similar criticisms about technical documents in general.

The purpose of this chapter is thus to bring together and describe the best-practice features of English language technical documents. However, it is not an exhaustive list of every best-practice feature. Rather, it is the linguistic and organisational best-practice features most often identified by the author in literature published over the last two and a half decades as helping make the information in modern technical documents understandable as this is the main focus of the controlled languages looked at in the previous chapter. The emphasis of this literature is on printed technical documents. However, the identified best-practice features are equally applicable to electronic technical documents.

The best-practice features have been identified from an extensive international body of academic, government and professional literature. This literature can be classified into three broad categories. The first category is guidelines from national and international government and professional regulatory and standards bodies such as the British Standards Institution, Council of the European Union, TCeurope and United States Consumer Product Safety Commission.

The second category is academic and professional literature related to technical document development. Over the past two and half decades, an extensive body of literature has emerged on the best-practice features of technical documents by writers such as Burnett, Hackos, Kirkman, Robinson and Schriver.

The third category is published style guides for technical documents. These provide best-practice guidelines for technical document development and include *Science and Technical Writing: A Manual of Style* (2001), *The Global English Style Guide* (2008), *Read Me First: A Style*

Guide for the Computer Industry (2009), *Microsoft Manual of Style* (2012) and *FranklinCovey Style Guide for Business and Technical Communication* (2012).

The chapter then concludes by considering some of the benefits to manufacturing companies of developing understandable technical documents.

Linguistic Best-Practice Features

Fig. 3.1 lists the linguistic best-practice features most often identified in the literature as helping make the information in modern technical documents understandable.

1. Uses short, simply structured sentences.
2. Uses the active voice, particularly in sentences containing instructions.
3. Uses the imperative mood, particularly in sentences containing instructions.
4. Uses mainly positively worded sentences. Uses negatively worded sentences wherever possible in cautions and warnings.
5. Does not use unnecessary words.
6. Does not use telegraphic language.
7. Uses simple verb tenses and forms, particularly the simple present tense. Avoids the gerund form.
8. Spells words consistently.
9. Uses punctuation consistently.
10. Limits use of technical terms.
11. Uses words consistently.
12. Limits use of abbreviations and contractions.
13. Does not use colloquialisms, idioms and region-specific words.
14. Does not use long strings of nouns and adjectives.

Fig. 3.1 Linguistic best-practice features

The first linguistic best-practice feature of modern technical documents most often identified in the literature is the use of short, simply structured sentences (for example, Crown 1992; Vesper 1997; Esselink 2000; Young 2002; Kirkman 2005; Lipus 2006; Alred et al. 2009; Sun Technical Publications 2009; Beisse 2013; Albert et al. 2014; Swisher 2014).

A long sentence generally contains multiple coordinating or subordinating clauses and thus requires readers to work hard to cognitively identify and process the information contained in it. This is especially true if the readers have limited English language and/or literacy skills. It is thus particularly important that manufacturing companies use short, simple sentences in English language technical documents for target readers with mixed English language and/or literacy skills. In this regard, Young (2002) and Wallace and Webber (2009) make clear that the use of short, simple sentences in technical documents does not carry with it an implied judgement on the intelligence of the target readers. In fact, it is evident from Chap. 1 that there has been a gradual historical movement towards the use of shorter, simpler sentences in technical documents.

The second linguistic best-practice feature of modern technical documents most often identified is the use of the active voice, particularly for instructional sentences (for example, Council of the European Union 1998; British Standards Institute 2001; Rubens 2001; Amador and Keller 2002; Barker 2003; United States Consumer Product Safety Commission 2003; Hargis et al. 2004; Lannon 2006; Alred et al. 2009; Robinson 2009; Microsoft Corporation 2012; Albert et al. 2014).

Information in the active voice is generally easier to cognitively identify and process than information in the passive voice as, first, sentences in the active voice are generally shorter and more direct than sentences in the passive voice and, second, sentences in the active voice make the readers the subject of the information.

The third related linguistic best-practice feature most often identified is the use of the imperative mood, particularly for instructional sentences (for example, British Standards Institute 1992; Eisenberg 1992; Reep 1996; Pringle and O'Keefe 2003; Hargis et al. 2004; Markel 2004; Campanizzi 2005; Alred et al. 2009; Cutts 2009; Sun Technical Publications 2009; Graves and Graves 2012; Microsoft Corporation 2012; Van Laan 2012).

The imperative mood is used to tell someone to do something, thus making it particularly suitable for the instructional sentences in technical documents. It is evident from Chap. 1 that the imperative mood, like the active voice, has historically been used for instructional sentences in technical documents.

The fourth most identified linguistic best-practice feature is a preponderance of positively worded sentences over negatively worded sentences (for example, Andrews 2001; Pearsall 2001; United States Consumer Product Safety Commission 2003; Markel 2004; Hargis et al. 2004; Burnett 2005; Kohl 2008; Sun Technical Publications 2009).

Positively worded information is generally easier to cognitively identify and process than negatively worded information as the latter necessitates readers first understanding what not to do and then, based on this understanding, inferring what to do. Furthermore, research by the United States Consumer Product Safety Commission (2003) suggests that the information in technical documents is more likely to be complied with if it is worded as what readers should do rather than as what readers should not do. However, turning to Burnett (2005) and Kohl (2008), one finds that the use of negatively worded sentences is, conversely, appropriate for the cautions and warnings in technical documents to linguistically differentiate them from the rest of the information.

The fifth linguistic best-practice feature most often identified is the use of concise language (for example, British Standards Institute 1993; Council of the European Union 1998; Andrews 2001; Amador and Keller 2002; Hargis et al. 2004; Burnett 2005; Kohl 2008; Martinez et al. 2008; Robinson 2009; Tyagi and Misra 2011; Beisse 2013; Albert et al. 2014).

Unnecessary words in technical documents can negatively affect understandability as they necessitate readers being able to identify and eliminate them. This can be particularly challenging for readers with limited English language and/or literacy skills. Strunk makes clear that "A sentence should contain no unnecessary words ... for the same reason that a drawing should have no unnecessary lines and a machine no unnecessary parts" (1999: 21).

The language is concise, but it is not telegraphic (for example, Esselink 2000; Kirkman 2001; Rubens 2001; Barker 2003; Markel 2004; Campanizzi 2005; Kohl 2008; Alred et al. 2009; Beisse 2013; Whitaker and Mancini 2013).

The use of telegraphic sentences in technical documents can negatively affect understandability as syntactic clues to meaning—such as articles (the, a, an), demonstrative adjectives (this, that, these, those) and relative pronouns (that, which, when)—are missed out.

The seventh linguistic best-practice feature most often identified is the use of simple verb tenses and forms and in particular the simple present tense (for example, Lee 1993; Weiss 2003; Hargis et al. 2004; Apple Inc. 2008; Alred et al. 2009; Sun Technical Publications 2009; Microsoft Corporation 2012).

By way of illustration, the *Microsoft Manual of Style* recommends avoiding the gerund form (Microsoft Corporation 2012). This is because the gerund and present continuous form of a verb have the same "ing" construction. This means that readers with limited English language skills may be unable to differentiate between them and thus be unable to cognitively identify and process the information they need.

The eighth most identified linguistic best-practice feature is the consistent spelling of words (for example, Council of the European Union 1998; Pringle and O'Keefe 2003; Hargis et al. 2004; Weiss 2005; Kohl 2008; Robinson 2009; Greenlaw 2012). Moreover, the same word can be spelled differently in different countries, thus necessitating technical communicators to choose one spelling. It is important for understandability that technical communicators are consistent with their spelling choices.

A related linguistic best-practice feature is the consistent use of punctuation (for example, Jones 2000; Pringle and O'Keefe 2003; Hargis et al. 2004; Sun Technical Publications 2009; Greenlaw 2012; Beisse 2013). It is important for technical document understandability that sentences containing similarly worded instructions are punctuated in a similar way.

The tenth linguistic best-practice feature most often identified is the limited use of technical terms (for example, Jones 1997; Council of the European Union 1998; Amador and Keller 2002; United States Consumer Product Safety Commission 2003; Hargis et al. 2004; TCeurope 2004; Microsoft Corporation 2012). In addition, technical terms, when used, are explained clearly and used consistently (for example, Eisenberg 1992; British Standards Institute 1993, 2001; Amador and Keller 2002; United States Consumer Product Safety Commission 2003; Hargis et al. 2004; TCeurope 2004; Byrne 2006; Sharoff and Hartley 2012; Covey 2012). Crystal points out that the information in modern technical documents is often difficult to understand because "many manufacturers assume that

3 Best-Practice Features of Modern Technical Documents

the user will know all the technical terms about their product, and do not bother to explain them" (On Read Rage, 2008).

In fact, all words should be used consistently with the same meaning to help make the information in technical documents understandable (for example, Council of the European Union 1998; Esselink 2000; Kirkman 2001; Rubens 2001; Amador and Keller 2002; Hargis et al. 2004; TCeurope 2004; Robinson 2009; Covey 2012; Swisher 2014).

The use of different words to mean the same thing, or the same word to mean different things, can negatively affect understandability as it necessitates readers being able to determine which meaning is intended from the context of the sentence. The first of these problems can be simply illustrated by considering the number of words that are often used interchangeably to denote depressing a key on a keyboard. A partial list might include hit, press, punch, strike, tap and touch.

Another linguistic best-practice feature is the limited use of abbreviations and contractions. In addition, the meaning of any abbreviation or contraction is explained the first time it is used (for example, British Standards Institute 1993; Council of the European Union 1998; Esselink 2000; Amador and Keller 2002; United States Consumer Product Safety Commission 2003; Lipus 2006; Kohl 2008).

Amador and Keller (2002) point out that the use of abbreviations and contractions can negatively affect the understandability of the information in technical documents for both native and non-native English users as the same abbreviations and contractions may not be used throughout a country.

A related linguistic best-practice feature most often identified in the literature is the absence of colloquialisms, idioms and region-specific words (for example, Esselink 2000; Kirkman 2001; Rubens 2001; Hargis et al. 2004; Weiss 2005; Kohl 2008; Martinez et al. 2008; Microsoft Corporation 2012; Swisher 2014). This is because readers with limited English language skills, and readers from a different region to the technical communicator, may not be able to derive their meaning from the context of the text and thus be unable to cognitively identify and process the information they need.

The final linguistic best-practice feature of modern technical documents most often identified in the literature is the absence of long strings of nouns or adjectives (for example, Pearsall 2001; Rubens 2001; Young 2002; United States Consumer Product Safety Commission 2003; Markel 2004; Hargis et al. 2004; Burnett 2005; Lipus 2006; Sun Technical Publications 2009; Covey 2012). Long word strings can negatively affect understandability because of the potential difficulty of understanding how each word in the string relates to the others.

In describing these linguistic best-practice features of modern technical documents, several references have been made to readers with limited English literacy skills. In concluding this section, it is important to point out that the readers of English language technical documents in countries with English as a native or official language (whether *de facto* or otherwise) can include readers with limited English literacy skills. The results of two surveys illustrate this point.

First, a nationwide study conducted by the US Department of Education in 2003 revealed that 22% (30 million) of American adults 16 years of age and older have only basic document literacy and 14% (19 million) have below basic document literacy. For the purposes of this study, the term document literacy referred to the ability to understand the meaning of non-continuous text such as that found in technical documents. (For interested readers, the key findings of the study can be viewed online at http://nces.ed.gov/naal/kf_demographics.asp.)

Second, a nationwide study conducted by the British Department for Education and Skills between 2002 and 2003 revealed that 16% (5.2 million) of British adults 16 years of age and older have a literacy level at or below the level expected of an 11-year-old. (For interested readers, the key findings of this study can be viewed online at http://webarchive.nationalarchives.gov.uk/20130123124929/http://www.education.gov.uk/aboutdfe/foi/disclosuresaboutchildrenyoungpeoplefamilies/a0065050/illiteracy-amongst-children-and-adults.)

It is widely argued (for example, Escoe 2001; United States Consumer Product Safety Commission 2003; Robinson 2009) that English language technical documents for a general readership of native English users should be at a 6th–8th US grade level of understandability (approximately 11–13 years old) and for a general readership of native and non-native English users at a slightly lower US grade level of understandability. There are a number of tests that have been developed to estimate technical document understandability in terms of the US grade level needed to understand them such as the Automated Readability Index and Flesch-Kincaid Grade Level Formula. The accuracy of these kinds of tests is questioned by some researchers (for example, Redish and Selzer 1985; Samson 1993; Sides 1999; Burnett 2005) as they are calculated based on only a limited number of the linguistic features of a technical document. Nevertheless, many other researchers (for example, Escoe 2001; DuBay 2007; Cutts 2009; Robinson 2009) view them as providing a useful starting point for estimating the understandability of a technical document.

Organisational Best-Practice Features

Fig. 3.2 lists the organisational best-practice features most often identified in the literature as helping make the information in modern technical documents understandable.

---

1. Uses parallel language structures, particularly for task-oriented headings and sentences containing instructions.
2. Uses a tabular structure for sentences, particularly for sentences containing instructions.
3. Numbers sentences containing consecutive instructions.
4. Limits each sentence to one piece of information or one instruction. (Exception: Puts simultaneous, or nearly simultaneous, instructions in the same sentence.)
5. Arranges sentences containing information (not instructions) into short paragraphs of no more than nine sentences with one topic per paragraph.
6. Arranges sentences containing instructions into task-oriented chunks.
7. Arranges task-oriented chunks in the order that end users are likely to need the information.
8. Uses task-oriented headings for task-oriented chunks.

Fig. 3.2 Organisational best-practice features

The first organisational best-practice feature of modern technical documents most often identified in the literature is the use of parallel language structures, particularly for task-oriented headings and instructional sentences (for example, Andrews 2001; Pearsall 2001; Rubens 2001; Markel

2004; Byrne 2006; Martinez et al. 2008; Robinson 2009; Sun Technical Publications 2009; Covey 2012; Microsoft Corporation 2012).

The use of parallel, as opposed to non-parallel, language structures—such as in the following example—can help make the information in technical documents understandable as, first, readers need only decode the structural meaning of the parallel headings and instructional sentences once and, second, it increases the predictability of the information.

Non-parallel language structure:

- Press the PLAY button to begin playback.
- Pressing the PAUSE button pauses playback.

Parallel language structure:

- Press the PLAY button to begin playback.
- Press the PAUSE button to pause playback.

It is evident from Chap. 1 that parallel language structures have been used for task-oriented headings in technical documents as far back as Chaucer's *A Treatise on the Astrolabe* (1391).

The second most identified organisational best-practice feature is the tabular structuring of sentences, particularly instructional sentences (for example, Debs 1988; British Standards Institute 1993; Robinson 2009; Whitaker and Mancini 2013). Closely related to this is the sequential numbering of sentences containing consecutive instructions (for example, Andrews 2001; McMurrey 2002; Pringle and O'Keefe 2003; United States Consumer Product Safety Commission 2003; Campanizzi 2005; Apple Inc. 2008; Sun Technical Publications 2009; Graves and Graves 2012).

The use of a numbered, tabular structure positively affects understandability as it helps the end users of a technical good visually scan and recognise information when looking backwards and forwards between the technical document and the technical good. This is of particular importance as, in the words of Robinson, "manuals are almost never read in isolation. No one curls up next to the fireplace on a rainy afternoon to read a manual. People almost always read manuals while they are interacting with the product" (2009: 54). In addition, it can make the instructional information appear less dense and thus less difficult to understand. The use of a numbered, tabular structure for instructional sentences was seen in Chap. 1 in the historical technical document for the Leica camera.

The fourth organisational best-practice feature most often identified is that each sentence is limited to one piece of information or one instruction (for example, British Standards Institute 2001; Young 2002; United States Consumer Product Safety Commission 2003; Campanizzi 2005; Lipus 2006; Martinez et al. 2008; Sun Technical Publications 2009; Graves and Graves 2012). This helps make technical documents understandable as it enables readers to more easily identify and process the piece of information or instruction in each sentence. However, an exception to this is when two or more instructions are simultaneous, or nearly simultaneous, rather than sequential. Two or more instructions are then included in the same sentence to emphasise, and facilitate understanding of, their concurrent (or nearly concurrent) nature.

The fifth most identified organisational best-practice feature is the chunking of sentences containing information (rather than instructions) into short paragraphs, with no more than one topic in each paragraph (for example, Burnett 2005; Campanizzi 2005; Martinez et al. 2008; Cutts 2009; Covey 2012; Graves and Graves 2012).

It is important for technical document understandability that long paragraphs with multiple topics are avoided as they can be particularly challenging for readers with limited English language and/or literacy skills. Horn (1998) has undertaken extensive research on what is termed the chunking principle. This refers to the chunking of information in documents into short sections of no more than seven plus or minus two sentences. Horn's research suggests that this method of organising information increases its understandability. The selection of the numbers seven and two can be traced back to memory research undertaken by the cognitive psychologist Miller (1956) that suggests humans can only retain seven plus or minus two items of information in their short-term memory.

A related best-practice feature of modern technical documents most often identified in the literature is the chunking of sentences containing instructions into task-oriented chunks, with each chunk containing instructions for completing a specific task (for example, Hackos 1994; Council of the European Union 1998; Mehlenbacher 2003; United States Consumer Product Safety Commission 2003; Pringle and O'Keefe 2003; Hargis et al. 2004; Graves and Graves 2012; Swisher 2014).

It is clear from Chap. 1 that this has historically been a feature of technical documents. In addition, it is important for understandability that the task-oriented chunks are organised in the order that the end users of the technical goods are likely to perform tasks with them and thus need the technical documents (for example, Gleason and Wackerman 2003; United States Consumer Product Safety Commission 2003; Hargis et al. 2004; TCeurope 2004).

The eighth and final organisational best-practice feature most often identified is the use of task-oriented headings for task-oriented chunks as they enable readers to easily and quickly identify and process the information in the chunks (for example, Krull 1994; Hackos 1994; Sims 2002; Barker 2003; Hargis et al. 2004; Campanizzi 2005).

In summary, this and the previous section have identified the most widely agreed upon linguistic and organisational best-practice features of modern technical documents that help make the information in them understandable. Having read this information, readers may be interested to know the key benefits of developing understandable technical documents.

Key Benefits of Developing Understandable Technical Documents

There is also much agreement in the literature on the key benefits for manufacturing companies of developing understandable technical documents.

The first agreed benefit is improved end user experiences (for example, Burgess 1984; Gingras 1987; British Standards Institute 1993; Hackos 1994). Developing understandable technical documents can allow the features and functions of technical goods to be used quicker, more fully and with less error.

The second related benefit is a reduction in technical support costs (for example, Gingras 1987; British Standards Institute 1993; Hackos 1994; Spencer and Yates 1995; Robinson and Etter 2000; Markel 2004; TCeurope 2004; Granered 2005; Robinson 2009). Manufacturer help desks and service centres are often a primary source of support for end users who cannot understand the information in technical documents. Developing technical documents that are understandable for the target readers can potentially reduce the expenses associated with providing these services.

The third agreed potential benefit is a reduced risk of physical injury to end users or damage to goods (for example, Gingras 1987; British Standards Institute 1993; Klauke 1994; Murphy et al. 1998; Warren 2002; Markel 2004).

The final agreed potential benefit is a reduced risk of litigation (for example, Gingras 1987; British Standards Institute 1993; Manning 1997; Murphy et al. 1998; Robinson and Etter 2000; Hartley and Paris 2001; Lipus 2006; Cutts 2009; Robinson 2009). There are many conventions, laws and resolutions that make reference to technical documents, as is illustrated in the following examples. Article 1 of The Hague Convention on the Law Applicable to Products Liability of October 2, 1973, states that manufacturers are liable for "damage in consequence of a misdescription of the product or of a failure to give adequate notice of its qualities, its characteristics or its method of use". (For interested readers, the full text can be viewed online at https://www.hcch.net/en/instruments/conventions/full-text/?cid=84.) The 1974 Trade Practices Act of the Parliament of Australia similarly states that "instructions for, or warnings about, doing or refraining from doing anything with or in relation to the product" are considered in determining if it is defective. (For interested readers, the full text can be viewed online at http://www.ruralandgeneral.co.nz/consumer-what-is-product-liability.php.)

Summary

It was shown in Chap. 2 that many controlled languages shape and constrain the linguistic and organisational features of technical documents to help make the information in them understandable. It was shown in this chapter that there is consistent agreement in an extensive international body of academic, government and professional literature on the linguistic and organisational best-practice features of modern technical documents that help make the information in them understandable.

This raises the question to be considered in Chap. 4 as to whether, and to what extent, the rule sets from the controlled languages introduced in Chap. 2 reflect the identified linguistic and organisational best-practice features introduced in this chapter.

Bibliography

Albert, Camille, Mathilde Janier, and Patrick Saint-Dizier. 2014. The Art of Writing Technical Documents. In *Challenges of Discourse Processing: The Case of Technical Documents*, ed. Patrick Saint-Dizier, 53–72. Cambridge: Cambridge Scholars Publishing.

Alred, Gerald J., Charles T. Brusaw, and Walter E. Oliu. 2009. *Handbook of Technical Writing*. 9th ed. New York: St. Martin's Press.

Amador, Mable, and Yvonne Keller. 2002. *International English Manual*. Los Alamos: Los Alamos National Laboratory.

Andrews, Deborah C. 2001. *Technical Communication in the Global Community*. 2nd ed. New Jersey: Prentice Hall.

Apple Inc. 2008. *Apple Publications Style Guide*. Cupertino: Apple Inc.

Barker, Thomas T. 2003. *Writing Software Documentation: A Task-Oriented Approach*. 2nd ed. New York: Longman.

Beisse, Fred. 2013. *A Guide to Computer User Support for Help Desk & Support Specialists*. 5th ed. Boston: Cengage Learning.

British Standards Institute. 1992. *BS 4884-1 Technical Manuals—Part 1: Specification for Presentation of Essential Information*. London: British Standards Institute.

———. 1993. *BS 4884-2 Technical Manuals—Part 2: Guide to Content*. London: British Standards Institute.

———. 2001. *BS EN 62079:2001 Preparation of Instructions—Structuring, Content and Presentation*. London: British Standards Institute.

Brockmann, R. John. 1990. *Writing Better Computer User Documentation: From Paper to Hypertext*. New York: John Wiley & Sons.

Burgess, John A. 1984. *Design Assurance for Engineers and Managers*. New York: Marcel Dekker, Inc.

Burnett, Rebecca E. 2005. *Technical Communication*. 6th ed. Massachusetts: Thomson Wadsworth.

Byrne, Jody. 2006. *Technical Translation: Usability Strategies for Translating*. Dordrecht: Springer.

———. 2012. *Scientific and Technical Translation Explained: A Nuts and Bolts Guide for Beginners*. London: Routledge.

Campanizzi, Jane. 2005. *Effective Writing for the Quality Professional; Creating Useful Letters, Reports and Procedures*. Milwaukee: ASQ Quality Press.

Chaucer, Geoffrey. [1391] 1870. *A Treatise on the Astrolabe*. London: John Russell Smith.

Cohen, Gerald. 1995. Graphics: The Missing Half of Technical Documentation. In *1995 IEEE International Professional Communication Conference: Smooth Sailing to the Future*, ed. Institute of Electrical and Electronics Engineers, 137–139. Savannah: Institute of Electrical and Electronics Engineers.

Council of the European Union. 1998. 'Resolution C411 (1998)' Council Resolution 98/C 411/01 of 17th December 1998 on Operating Instructions for Technical Consumer Goods. *EUR-Lex*. http://eur-lex.europa.eu/LexUriServ/LexUriServ.do?uri=OJ:C:1998:411:0001:0004:EN:PDF. Accessed 8 August 2016.

Covey, Stephen R. 2012. *FranklinCovey Style Guide for Business and Technical Communication*. 5th ed. Salt Lake City: FranklinCovey.

Crown, James. 1992. *Effective Computer User Documentation*. New York: Van Nostrand Reinhold.

Crystal, David. 2008. On Read Rage. *DCBlog*. http://david-crystal.blogspot.co.uk/2008/11/on-read-rage.html. Accessed 24 August 2016.

Cutts, Martin. 2009. *Oxford Guide to Plain English*. 3rd ed. Oxford: Oxford University Press.

Debs, Mary B. 1988. A History of Advice: What Experts Have to Tell Us. In *Effective Documentation: What We Have Learned from Research*, ed. Stephen Doheny-Farina, 11–23. Cambridge: MIT Press.

DuBay, William H. 2007. *Smart Language: Readers, Readability, and the Grading of Text*. Costa Mesa: Impact Information.

Eisenberg, Anne. 1992. *Effective Technical Communication*. 2nd ed. New York: McGraw-Hill.

Escoe, Adrienne. 2001. *The Practical Guide to People-Friendly Documentation*. 2nd ed. Milwaukee: ASQ Quality Press.

Esselink, Bert. 2000. *A Practical Guide to Localization*. Amsterdam: John Benjamins Publishing Company.

Gingras, Becky. 1987. Simplified English in Maintenance Manuals. *Technical Communication* First Quarter: 24–28.
Gleason, James P., and Joan P. Wackerman. 2003. Manual Dexterity-What Makes Instruction Manuals Usable. In *Writing and Speaking in the Technology Professions: A Practical Guide*, 2nd ed., ed. David Beer, 158–160. New York: John Wiley & Sons.
Granered, Erik. 2005. *Achieving Outstanding Customer Service Across Cultures & Time Zones*. Boston: Nicholas Brealey International.
Graves, Heather, and Roger Graves. 2012. *A Strategic Guide to Technical Communication*. 2nd ed. Peterborough: Broadview Press.
Greenlaw, Raymond. 2012. *Technical Writing, Presentation Skills, and Online Communication: Professional Tools and Insights*. Hershey: Information Science Reference.
Griffiths, Dave. 2001. Design for Usability—No One Reads Manuals. In *The ISTC Handbook of Professional Communication and Information Design*, ed. The Institute of Scientific and Technical Communicators, 1–12. Peterborough: The Institute of Scientific and Technical Communicators.
Hackos, JoAnn T. 1994. *Managing Your Documentation Projects*. New York: John Wiley & Sons.
Hargis, Gretchen, Michelle Carey, Ann K. Hernandez, Polly Hughes, Deirdre Longo, Shannon Rouiller, and Elizabeth Wilde. 2004. *Developing Quality Technical Information: A Handbook for Writers and Editors*. 2nd ed. New Jersey: IBM Press.
Hartley, Anthony, and Cécile Paris. 2001. Translation, Controlled Languages, Generation. In *Exploring Translation and Multilingual Text Production: Beyond Context*, ed. Erich Steiner and Colin Yallop, 307–326. Berlin: Walter de Gruyter.
Horn, Robert E. 1998. Structured Writing as a Paradigm. In *Instructional Development: State of the Art*, ed. Alexander Romiszowski and Charles Dills. Englewood Cliffs: Educational Technology Publications.
Jones, Dan. 1997. *Technical Writing Style*. Upper Saddle River: Pearson Education.
———. 2000. *The Technical Communicator's Handbook*. Boston: Allyn and Bacon.
Kirkman, John. 2001. Choosing Language for Effective Technical Writing. In *The ISTC Handbook of Professional Communication and Information Design*, ed. The Institute of Scientific and Technical Communicators, 59–76. Peterborough: The Institute of Scientific and Technical Communicators.

———. 2005. *Good Style: Writing for Science and Technology*. 2nd ed. London: Routledge.

Klauke, Michael. 1994. National Standards—Their Impact on Text Production and Quality. In *Quality of Technical Documentation*, ed. Michaël Steehouder, Carel Jansen, Pieter van der Poort, and Ron Verheijen, 161–170. Amsterdam: Rodopi.

Kohl, John R. 2008. *The Global English Style Guide*. Cary: SAS Institute Inc.

Krull, Robert. 1994. Comparative Assessment of Document Usability with Writing Quality Measures. In *STC Annual Conference Proceedings*, ed. Society for Technical Communication, 216–218. Arlington: Society for Technical Communication.

Lannon, John M. 2006. *Technical Communication*. 10th ed. New York: Pearson Education.

Lee, Arthur. 1993. Controlled English with and Without Machine Translation. In *Translating and the Computer 15*, ed. The Association for Information Management, 35–39. London: Aslib.

Lipus, Teresa. 2006. International Consumer Protection: Writing Adequate Instructions for Global Audiences. *Journal of Technical Writing and Communication* 36(1): 75–91.

Manning, Michael D. 1997. Hazard Communication 101 for Technical Writers. In *STC Annual Conference Proceedings*, ed. Society for Technical Communication, 446–449. Arlington: Society for Technical Communication.

Markel, Mike. 2004. *Technical Communication*. 7th ed. Boston: Bedford/St. Martin's.

Martin, Michelle (Producer). August 21, 2009. *How to Write an Instruction Manual*. London: BBC Radio 4.

Martinez, Diane, Tanya Peterson, Carrie Wells, Carrie Hannigan, and Carolyn Stevenson. 2008. *Kaplan Technical Writing: A Resource for Technical Writers at All Levels*. New York: Kaplan Publishing.

McMurrey, David A. 2002. *Power Tools for Technical Communication*. Boston: Wadsworth.

Mehlenbacher, Brad. 2003. Documentation: Not Yet Implemented, But Coming Soon. In *The Human-Computer Interaction Handbook: Fundamentals, Evolving Technologies and Emerging Applications*, ed. Andrew Sears and Julie A. Jacko, 527–543. Mahwah: Lawrence Erlbaum Associates.

Microsoft Corporation. 2012. *Microsoft Manual of Style*. 4th ed. Redmond: Microsoft Press.

Miller, George A. 1956. The Magical Number Seven, Plus or Minus Two: Some Limits on Our Capacity for Processing Information. *Psychological Review* 63(1): 81–96.

Murphy, Dawn, Jane Mason, and Stuart Sklair. 1998. Improving Translation at the Source. In *Translating and the Computer 20*, ed. The Association for Information Management. London: Aslib.

Pearsall, Thomas E. 2001. *The Elements of Technical Writing*. 2nd ed. Needham Heights: Allyn & Bacon.

Pringle, Alan S., and Sarah S. O'Keefe. 2003. *Technical Writing 101: A Real-World Guide to Planning and Writing Technical Documentation*. 2nd ed. Dallas: Scriptorium Press.

Redish, Janice C., and Jack Selzer. 1985. The Place of Readability Formulas in Technical Communication. *Technical Communication* 32(4): 46–52.

Reep, Diana C. 1996. *Technical Writing: Principles, Strategies, and Readings*. 3rd ed. Boston: Allyn & Bacon.

Robinson, Patricia A. 2009. *Writing and Designing Manuals and Warnings*. 4th ed. Boca Raton: CRC Press.

Robinson, Patricia A., and Ryn Etter. 2000. *Writing and Designing Manuals*. 3rd ed. Boca Raton: CRC Press.

Rubens, Philip. 2001. *Science and Technical Writing: A Manual of Style*. 2nd ed. Abingdon: Routledge.

Samson, Donald C. 1993. *Editing Technical Writing*. Oxford: Oxford University Press.

Schriver, Karen A. 1997. *Dynamics in Document Design*. New York: John Wiley & Sons, Inc.

Sharoff, Serge, and Anthony Hartley. 2012. Lexicography, Terminology and Ontologies. In *Handbook of Technical Communication*, ed. Alexander Mehler, Laurent Romary, and Dafydd Gibbon, 317–346. Berlin: Walter de Gruyter GmbH.

Sides, Charles H. 1999. *How to Write and Present Technical Information*. 3rd ed. Cambridge: Cambridge University Press.

Sims, Brenda R. 2002. *Technical Communication for Readers and Writers*. 2nd ed. Boston: Houghton Mifflin.

Spencer, Cathy J., and Diana K. Yates. 1995. A Good User's Guide Means Fewer Support Calls and Lower Support Costs. *Technical Communication* 42(1): 52–55.

Strunk, William. 1999. *The Elements of Style*. 4th ed. New York: Longman.

Sun Technical Publications. 2009. *Read Me First! A Style Guide for the Computer Industry*. 3rd ed. Palo Alto: Prentice Hall.

Swisher, Val. 2014. *Global Content Strategy: A Primer*. Laguna Hills: XML Press.

TCeurope. 2004. Usable and Safe Operating Manuals for Consumer Goods: A Guideline. Version 1.0. *COM&TEC*. http://www.comtec-ct.org/images/PDF/SecureDoc_EN.PDF. Accessed 8 August 2016.

Thimbleby, Harold. 2000. On Discerning Users. In *How to Make User Centred Design Usable. Technical Report TRITA-NA-D0006, CID-72*, ed. Jan Gulliksen, Ann Lantz, and Inger Boivie, 63–78. Stockholm: Royal Institute of Technology.

Tyagi, Kavita, and Padma Misra. 2011. *Basic Technical Communication*. New Delhi: PHI Learning Private Limited.

United States Consumer Product Safety Commission. 2003. Manufacturer's Guide to Developing Consumer Product Instructions. *Consumer Product Safety Commission*. https://www.cpsc.gov/PageFiles/103077/guide.pdf. Accessed 8 August 2016.

Van Laan, Krista. 2012. *The Insider's Guide to Technical Writing*. Arlington: Society for Technical Communication.

Vesper, James L. 1997. *Documentation Systems: Clear and Simple*. Boca Raton: CRC Press.

Wallace, Michael, and Larry Webber. 2009. *IT Governance Policies & Procedures*. New York: Aspen Publishers.

Warren, Thomas L. 2001. Communicating Style Rules to Editors of International Standards: An Analysis of ISO TC 184/SC4 Style Documents. *Journal of Technical Writing and Communication* 31(2): 159–173.

———. 2002. Cultural Influences on Technical Manuals. *Journal of Technical Writing and Communication* 32(2): 111–123.

Weiss, H. Eugene. 2003. *Chrysler, Ford, Durant, and Sloan: Founding Giants of the American Automotive Industry*. Jefferson: McFarland.

Weiss, Edmond H. 2005. *The Elements of International English Style*. New York: M.E. Sharpe.

Whitaker, Jerry C., and Robert K. Mancini. 2013. *Technical Documentation and Process*. Boca Raton: CRC Press.

Young, Matt. 2002. *The Technical Writer's Handbook*. 2nd ed. Sausalito: University Science Books.

4

Analysing Existing Controlled Languages Against the Best-Practice Features

Abstract The chapter first summarises the findings from existing controlled language analyses. It then brings together the information from the previous two chapters through analysing the controlled language rules introduced in the second chapter against the best-practice features introduced in the third chapter to ascertain if and how the rules reflect the best-practice features. The findings from the analysis help in identifying rules for a controlled language rule set model.

Keywords Analysing controlled languages • Controlled language rules • O'Brien (2003) • Schreurs and Adriaens (1992)

Introduction

To date, there have been few controlled language analyses published in the literature. This may surprise readers given that the first controlled language for English language technical documents, CFE, was developed over four decades ago. However, it was pointed out in Chap. 1 that a general lack of detailed information about controlled languages persists to this day.

A review of controlled language literature reveals just two analyses of multiple controlled languages by Schreurs and Adriaens of the Katholieke Universiteit Leuven (1992) and O'Brien of Dublin City University (2003). This chapter thus first summarises the findings from these existing analyses. It next introduces a new analysis. This compares the controlled language rule sets introduced in Chap. 2 with the linguistic and organisational best-practice features of modern technical documents introduced in Chap. 3 in order to ascertain whether, and to what extent, the rules used by manufacturing companies reflect the linguistic and organisational best-practice features identified from the literature. The findings from this analysis help in identifying new non-manufacturer specific rules that are used in the next chapter when developing a model controlled language rule set.

Existing Analyses of Controlled Language Rule Sets

It should be noted from the outset that the analyses undertaken by Schreurs and Adriaens (1992) and O'Brien (2003) differ from the analysis undertaken in this chapter in terms of their objectives, methodology and rigour. Nevertheless, their combined findings provide a partial, preliminary understanding of how existing controlled language rules reflect the best-practice features.

With regard to rigour, Schreurs and Adriaens list three controlled languages as being analysed, of which one is an "IBM manual" (1992: 208). This IBM manual is described as "in-house Information Development Guidelines on content, on-line information, style and vocabularies for customer and service information" (1992: 207). O'Brien lists eight controlled languages as being analysed, of which one is "IBM's EasyEnglish" (2003: 105). Reference to Bernth, one of the main developers of IBM's EasyEnglish, reveals that it is a "grammar checker++ ... standard grammar checking facilities such as spell-checking, word count (sentence length), and detection of passive constructions are available in addition to checks for ambiguity" (1997: 159). Moreover, a patent for IBM's EasyEnglish describes it as a "system … for analyzing English text and flagging possible errors and ambiguity". (For interested readers, the full patent can be viewed online at http://www.google.com/patents?id=kz8bAAAAEBAJ&printsec=abstract&zoom=4&source=gbs_overview_r&cad=0#v=onepage&q=&f=false.) These descriptions of IBM's manual and EasyEnglish suggest that they are not strictly controlled languages.

Notwithstanding, Schreurs and Adriaens' first and second pertinent findings are that the three controlled languages "contain a controlled vocabulary list" (1992: 209) and that "the spelling of words must be (sic) conform to the spelling used in the word lists" (1992: 209). It was seen in Chap. 2 that manufacturing companies use controlled vocabularies to help ensure that words are spelled and used uniformly in and across their technical documents. This language control reflects and is consistent with the linguistic best-practice features identified in Chap. 3 that words are spelled and used consistently in technical documents to help make the information in them understandable.

Their third pertinent finding is that the three controlled languages "allow noun clusters or compounds, if the number of nouns making up the compound does not exceed three" (1992: 210). This language control reflects and is consistent with the linguistic best-practice feature that long strings of nouns are not used in technical documents to help make the information understandable.

Schreurs and Adriaens' final pertinent finding is that the three controlled languages "control the verb forms to be used" (1992: 210). However, they do not provide sufficiently detailed information about this language control in their paper to determine the extent that it reflects and is consistent with the linguistic best-practice feature that simple verb tenses and forms, and in particular the simple present tense, are used in technical documents to help make the information in them understandable.

Before next considering O'Brien's findings, it should be noted that O'Brien classifies the majority of the analysed controlled languages as being machine-oriented. Researchers such as Nyberg et al. (2003), O'Brien (2003) and Reuther (2003) define machine-oriented controlled languages as being developed to help make the information in technical documents more translatable, particularly with machine translation systems. Nevertheless, it was seen in Chap. 2 that many controlled languages use the same simplified set of rules and controlled vocabulary to help make the information in technical documents understandable and aid its translation with machine translation systems. O'Brien's findings are thus pertinent to this chapter.

O'Brien's first pertinent finding is that the rule "'Keep procedural sentences as short as possible (20 words maximum)' is echoed in different ways by *all* CLs" (2003: 110), where CL is an abbreviation for controlled language. This language control reflects and is consistent with the linguistic best-practice feature identified in Chap. 3 that short sentences are used in technical documents to help make the information in them understandable.

O'Brien's second pertinent finding is that seven of the eight analysed controlled languages share the rule "Use the active voice" (2003: 110). This language control reflects and is consistent with the linguistic best-practice feature that the active voice is used to help make the information in technical documents understandable.

O'Brien's third pertinent finding is that seven of the eight analysed controlled languages share the rule "When appropriate, use an article (the, a, an) or a demonstrative adjective (this, these) before a noun" (2003: 110). This language control reflects and is consistent with the linguistic best-practice feature that telegraphic language is not used to help make the information understandable.

O'Brien's fourth pertinent finding is that "Six CLs share a rule regarding the use of the gerund, or, more specifically, they recommend avoiding it" (2003: 110). This language control reflects and is consistent with the linguistic best-practice feature that the gerund form is avoided to help make the information understandable.

O'Brien's fifth pertinent finding is that six of the analysed controlled languages share the rule "Do not make noun clusters of more than three nouns" (2003: 110). This language control reflects and is consistent with the linguistic best-practice feature that long noun strings are not used to help make the information understandable. In addition, this finding corresponds with the third pertinent finding of Schreurs and Adriaens.

O'Brien's sixth pertinent finding is that four of the analysed controlled languages share the rule "Make your instructions as specific as possible"

(2003: 110). This language control reflects and is consistent with the linguistic best-practice feature that unnecessary words are not used to help make the information understandable.

The final pertinent finding is that four of the analysed controlled languages share the rule "Use approved words from the Dictionary etc." (O'Brien 2003: 110). This language control reflects and is consistent with the linguistic best-practice feature that words are used consistently to help make the information in technical documents understandable. In addition, this finding is similar to the first and second pertinent findings of Schreurs and Adriaens.

These findings show that existing controlled language rules reflect and are consistent with at least seven of the best-practice features of modern technical documents introduced in Chap. 3. These are as follows:

1. Uses short sentences.
2. Uses the active voice.
3. Does not use telegraphic language.
4. Avoids the gerund form.
5. Does not use long strings of nouns.
6. Does not use unnecessary words.
7. Uses words consistently.

Nevertheless, the combined findings from both analyses are too limited in number for a comprehensive understanding of the relationship between existing controlled language rules and the identified best-practice features. The next section thus introduces and describes a more focused controlled language analysis.

New Analysis of Controlled Language Rule Sets

The purpose of the analysis is to ascertain whether, and to what extent, the CFE (Caterpillar), PACE (Perkins), EE (Ericsson), NSE (Nortel) and BCE (Bull) rule sets used primarily to help make the information in technical documents understandable reflect the linguistic and organisational best-practice features identified from the literature as helping make the information in technical documents understandable. The findings from the analysis help in identifying new non-manufacturer specific rules that will be used in Chap. 5 when developing a model controlled language rule set for technical documents.

For the convenience of readers, Fig. 4.1 lists the 22 best-practice features introduced in Chap. 3. Before looking at the findings, it should be

Linguistic best-practice features	Organisational best-practice features
• Uses short, simply structured sentences. • Uses the active voice, particularly in sentences containing instructions. • Uses the imperative mood, particularly in sentences containing instructions. • Uses mainly positively worded sentences. Uses negatively worded sentences wherever possible in cautions and warnings. • Does not use unnecessary words. • Does not use telegraphic language. • Uses simple verb tenses and forms, particularly the simple present tense. Avoids the gerund form. • Spells words consistently. • Uses punctuation consistently. • Limits use of technical terms. • Uses words consistently. • Limits use of abbreviations and contractions. • Does not use colloquialisms, idioms and region-specific words. • Does not use long strings of nouns and adjectives.	• Uses parallel language structures, particularly for task-oriented headings and sentences containing instructions. • Uses a tabular structure for sentences, particularly for sentences containing instructions. • Numbers sentences containing consecutive instructions. • Limits each sentence to one piece of information or one instruction. (Exception: Puts simultaneous, or nearly simultaneous, instructions in the same sentence.) • Arranges sentences containing information (not instructions) into short paragraphs of no more than nine sentences with one topic per paragraph. • Arranges sentences containing instructions into task-oriented chunks. • Arranges task-oriented chunks in the order that end users are likely to need the information. • Uses task-oriented headings for task-oriented chunks.

Fig. 4.1 Analysed best-practice features

4 Analysing Existing Controlled Languages... 75

noted that a small number of the rules from CFE, PACE, EE, NSE and BCE reflect more than one of these best-practice features.

The first finding is that all the controlled languages have one or more syntactic rules addressing sentence length and/or structure. All the rules reflect and are consistent with the linguistic best-practice feature listed in Fig. 4.1 that short, simply structured sentences are used in technical documents to help make the information in them understandable and, therefore, provide an informed basis from which to develop new non-manufacturer specific controlled language rules addressing sentence length and structure. In addition, this finding echoes the first pertinent finding from O'Brien's analysis. The syntactic rules that control sentence length are as follows:

- **CFE (Caterpillar):** Avoid long and complicated sentences. (rule 2)
- **PACE (Perkins):** Keep sentences short. (rule 1)
- **EE (Ericsson):** Write short sentences dealing with just one idea. (rule 15)
- **NSE (Nortel):** Write short sentences. Write sentences that have less than 22 words. (rule 2)
- **BCE (Bull):** Keep sentence length to a maximum of 25 words. (rule 2)

There is some variation in the specificity of the control, with three controlled languages (CFE, PACE and EE) stipulating the use of short sentences without specifying a fixed limit on the number of words in a sentence and two controlled languages (NSE and BCE) specifying a fixed limit of up to 25 words in a sentence. The establishment of a specific word limit imposes a greater degree of control over technical communicators and thus would seem to be particularly appropriate to promote consistency in and across technical documents.

The syntactic rules that control sentence structure are as follows:

- **BCE (Bull):** Use simple sentence structures. (rule 5)
- **NSE (Nortel):** Write sentences with simple sentence structure. (rule 3)
- **CFE (Caterpillar):** Avoid long and complicated sentences. (rule 2)
- **PACE (Perkins):** Don't change constructions in mid-sentence. (rule 4)

The specific wording of the rules varies slightly from one manufacturing company to another. However, they all share the same purpose of controlling sentence structure.

Two new non-manufacturer specific controlled language rules can be developed from these existing controlled language rules addressing sentence length and/or structure and the aforementioned best-practice feature. These are the syntactic rules "Use short sentences. (Limit sentences to no more than 25 words.)" and "Use simple sentence structures.". In the first rule, a specific limit has been placed on the number of words allowed in a sentence to promote consistency in and across technical documents.

The second finding of the analysis is that four of the controlled languages have a lexical rule addressing noun and/or adjective strings.

- **CFE (Caterpillar):** Avoid too many successive adjectives and nouns. (rule 4)
- **PACE (Perkins):** Avoid strings of nouns. (rule 9)
- **EE (Ericsson):** Do not make long strings of nouns or adjectives before a noun. (rule 2)
- **NSE (Nortel):** Avoid noun strings of more than three words. (rule 10)

This finding strongly correlates with the third pertinent finding from Schreurs and Adriaens' analysis and the fifth pertinent finding from O'Brien's analysis. All the rules reflect and are consistent with the linguistic best-practice feature from Fig. 4.1 that long strings of nouns and adjectives are not used to help make the information in technical documents understandable and, therefore, provide an informed basis from which to develop a new non-manufacturer specific controlled language rule addressing word strings.

There is again some variation in the specificity of the control, with three controlled languages (CFE, PACE and EE) not specifying any fixed numerical limit and one controlled language (NSE) specifying a fixed numerical limit of up to three words.

The new non-manufacturer specific controlled language rule "Limit word strings to no more than three nouns or adjectives." can be developed from these existing controlled language rules and the aforementioned best-practice feature. A specific limit has again been placed on the number of words allowed to promote consistency in and across technical documents.

The third finding of the analysis is that four of the controlled languages have a rule addressing the information load of sentences.

- **CFE (Caterpillar):** Avoid too many subjects in one sentence. (rule 3)
- **EE (Ericsson):** Write short sentences dealing with just one idea. (rule 15)
- **NSE (Nortel):** Write simple sentences that have one thought per sentence. (rule 3)
- **BCE (Bull):** One thought per sentence. (rule 4)

All the rules reflect and are consistent with the organisational best-practice feature listed in Fig. 4.1 that each sentence is limited to one piece of information or one instruction in technical documents to help make the information in it understandable and, therefore, provide an informed basis from which to develop a new non-manufacturer specific controlled language rule addressing information load.

Lee (1993), a former manager at Bull, reports that adherence to the BCE rule "One thought per sentence." (rule 4) helped make the information in Bull's technical documents understandable. However, it was noted in the previous chapter that two or more instructions should be included in the same sentence in technical documents if they are to be performed at, or almost at, the same time to facilitate understanding of their concurrent, or nearly concurrent, nature.

Three new non-manufacturer specific controlled language rules can be developed from these existing controlled language rules and the aforementioned best-practice feature. The rules "Limit each sentence to one instruction." and "Put two or more instructions in the same sentence if the end user is required to perform them at or almost at the same time." control the information load of sentences containing instructions, and the rule "Limit each sentence to one piece of information." controls the information load of sentences containing information (rather than instructions).

The description of the rule "Write short sentences dealing with just one idea." (rule 15) in the EE guide states that "In EE we try to write short sentences dealing with just one idea. Therefore punctuation is simple. But it must be used correctly and consistently. Its purpose is always the same: ... [making] ... understanding easier" (Telefonaktiebolaget LM Ericsson 1983: 18). This leads to the fourth finding, which is that three of the controlled languages have a grammatical rule that reflects the linguistic best-practice feature from Fig. 4.1 that punctuation is used consistently in technical documents to help make the information in them understandable.

- **CFE (Caterpillar):** Use punctuation correctly. (rule 9)
- **EE (Ericsson):** Use correct and consistent punctuation. (rule 17)
- **BCE (Bull):** Use correct punctuation. (rule 9)

Reference again to Lee (1993) reveals that the BCE rule "Use correct punctuation." (rule 9) was created because inconsistent use of punctuation was negatively affecting the ability of the end users of Bull's computers and consumer electronics goods to understand the information in the technical documents.

The new non-manufacturer specific controlled language rule "Use punctuation consistently. Always punctuate sentences containing similarly worded instructions in a similar way." can be developed from these existing controlled language rules and the aforementioned best-practice feature.

The fifth finding is that four of the controlled languages have one or more lexical rules addressing the use of abbreviations and/or contractions.

- **CFE (Caterpillar):** Avoid abbreviations, contractions and colloquialisms. (rule 8)
- **EE (Ericsson):** Do not use conversational expressions (idioms, contractions, slang). (rule 4)
  Do not use unnecessary abbreviations. (rule 5)
- **BCE (Bull):** Avoid abbreviations and colloquialisms. (rule 8)
- **NSE (Nortel):** Use only NSE approved abbreviations and acronyms. Use only abbreviations and acronyms that are in the NSE dictionaries. (rule 11)

All the rules reflect and are consistent with the linguistic best-practice feature in Fig. 4.1 that the use of abbreviations and contractions is limited to help make the information in technical documents understandable. They, therefore, provide an informed basis from which to develop a new non-manufacturer specific controlled language rule addressing abbreviation and contraction usage.

The CFE, EE and BCE rules require technical communicators to avoid abbreviations and/or contractions and the NSE rule requires technical communicators to use only approved abbreviations. The new non-manufacturer specific controlled language rule "Avoid using abbreviations and contractions unless they are part of the controlled vocabulary." can be developed from these existing controlled language rules and the aforementioned best-practice feature.

The sixth finding is that CFE, BCE and EE have a lexical rule requiring technical communicators to avoid colloquialisms or conversational expressions, EE and NSE have a lexical rule requiring technical communicators to avoid idioms and NSE has a lexical rule requiring technical communicators to avoid region-specific words. These rules are as follows:

- **CFE (Caterpillar):** Avoid abbreviations, contractions and colloquialisms. (rule 8)
- **BCE (Bull):** Avoid abbreviations and colloquialisms. (rule 8)
- **EE (Ericsson):** Do not use conversational expressions (idioms, contractions, slang). (rule 4)
- **NSE (Nortel):** Do not use jargon and idiomatic expressions. Use only terms that are in the NSE dictionaries. (rule 13)
  Use appropriate language. Avoid the following types of words: Words specific to a region or dialect (rule 14)

These rules, taken together, reflect and are consistent with the linguistic best-practice feature from Fig. 4.1 that the use of colloquialisms, idioms and region-specific words is avoided in technical documents to help make the information understandable. They thus provide an informed basis from which to develop the new non-manufacturer specific controlled language rule "Avoid using colloquialisms, idioms and region-specific words.".

The seventh finding is that four of the controlled languages have a stylistic rule that sentences should be positively worded.

- **CFE (Caterpillar):** Make positive statements. (rule 1)
- **BCE (Bull):** Make positive statements. (rule 1)
- **NSE (Nortel):** Write positive statements. (rule 1)
- **EE (Ericsson):** Avoid unnecessary negatives. (rule 11)

All these rules reflect and are consistent with the linguistic best-practice feature listed in Fig. 4.1 that sentences are predominantly positively worded in technical documents to help make the information understandable. It is interesting to note that this is the first rule in CFE, BCE and NSE. This suggests its perceived efficacy by Caterpillar, Bull and Nortel as a means to help make information understandable. In addition, the NSE rule has a qualifying subclause, listed in Chap. 2 of this book, that states "Exception: Use negative statements in warnings and cautions.". This exception is consistent with the best-practice feature that cautions and warnings are worded negatively to linguistically differentiate and draw attention to their content.

Two new non-manufacturer specific controlled language rules can be developed from this finding. The stylistic rule "Use positively worded sentences wherever possible." can be developed from the CFE, BCE, NSE and EE rules and the best-practice feature that sentences are predominantly positively worded. In addition, the stylistic rule "Use negatively worded sentences wherever possible in cautions and warnings to draw attention to their content." can be developed from the NSE subclause and the best-practice feature that cautions and warnings are worded negatively.

The eighth finding is that three of the controlled languages have a grammatical rule that the active voice should be used. This finding echoes the second pertinent finding from O'Brien's analysis.

- **EE (Ericsson):** Use only the active voice of verbs. (rule 8)
- **NSE (Nortel):** Use the active voice. Use the active voice instead of the passive voice. (rule 6)
- **BCE (Bull):** Avoid the passive voice. (rule 1)

All the rules reflect and are consistent with the linguistic best-practice feature from Fig. 4.1 that the active voice is used in technical documents, particularly in the instructional sentences, to help make the information understandable. They thus provide an informed basis from which to develop the new non-manufacturer specific controlled language rules "Use the active voice wherever possible." and "Always use the active voice in sentences containing instructions.".

The rationale behind the inclusion of the BCE rule "Avoid the passive voice." (rule 1) is given by Bull's Lee as "the Technical Author should be living in a permanent, active, present" (1993: 36). Lee's rationale also leads to the ninth finding, which is that four of the controlled languages have a grammatical rule addressing verb tense usage.

- **NSE (Nortel):** Use the present tense instead of the future tense. (rule 9)
- **EE (Ericsson):** Use only the simple past, simple present and simple future tenses. Always use the present tense when possible. (rule 9)
- **BCE (Bull):** Avoid the future tense. (rule 1)
- **CFE (Caterpillar):** Avoid complicated past and future tenses. (rule 6)

All the rules reflect and are consistent with the linguistic best-practice feature in Fig. 4.1 that simple verb tenses and forms, and in particular the simple present tense, are used in technical documents to help make the information understandable.

The new non-manufacturer specific controlled language rule "Use only the simple past, simple present, infinitive and simple future. Use the simple present wherever possible." can be developed from these existing controlled language rules and the aforementioned best-practice feature.

A related finding is that three of the controlled languages have a grammatical rule to avoid the gerund form. This finding correlates with the fourth pertinent finding from O'Brien's analysis.

- **PACE (Perkins):** Do not use -ing unless the word appears thus in the PACE dictionary. (rule 10)
- **EE (Ericsson):** "-ing" words are nouns that describe activities. (rule 16)
- **NSE (Nortel):** Do not use a gerund at the beginning of a sentence or in an ambiguous construction. (rule 7)

All the rules reflect and are consistent with the linguistic best-practice feature from Fig. 4.1 that use of the gerund form is avoided to help make the information in technical documents understandable. The new non-manufacturer specific controlled language rule "Avoid using gerunds unless they are part of the controlled vocabulary." can be developed from these existing controlled language rules and the aforementioned best-practice feature.

The eleventh finding is that three of the controlled languages have a syntactic rule that parallel language structures should be used.

- **CFE (Caterpillar):** Use uniform sentence structures. (rule 5)
- **NSE (Nortel):** Use parallel structure. (rule 5)
- **BCE (Bull):** Use parallel construction. (rule 6)

The rules reflect and are consistent with the organisational best-practice feature in Fig. 4.1 that parallel language structures are used in technical documents to help make the information understandable. The new non-manufacturer specific controlled language rule "Use parallel language structures wherever possible, particularly for task-oriented headings and sentences containing instructions." can be developed from these three existing controlled language rules and the aforementioned best-practice feature.

The twelfth finding is that PACE and EE have a stylistic rule that unnecessary words should not be used.

- **PACE (Perkins):** Omit redundant words. (rule 2)
- **EE (Ericsson):** Write clearly and simply. (rule 7)

This finding echoes the sixth pertinent finding from O'Brien's analysis. In addition, the rules reflect and are consistent with the linguistic best-practice feature that unnecessary words are not used to help make the information in technical documents understandable.

A closely related finding is NSE and PACE have a grammatical rule addressing telegraphic language.

- **NSE (Nortel):** Use a determiner with nouns and noun phrases when possible. (rule 15)
- **PACE (Perkins):** Don't omit conjunctions or relatives. (rule 7)

This finding echoes the third pertinent finding from O'Brien's analysis. In addition, the rules both reflect and are consistent with the linguistic best-practice feature that telegraphic language is not used in technical documents to help make the information understandable.

Two new non-manufacturer specific controlled language rules can be developed from these four existing controlled language rules and the aforementioned best-practice features. These are the stylistic rules "Avoid unnecessary words in sentences." and "Do not omit articles (the, a, an), demonstrative adjectives (this, that, these, those) or relative pronouns (that, which, when) from sentences.".

The thirteenth finding is that BCE and NSE have a lexical rule addressing the use of technical terms.

- **BCE (Bull):** Use valid terminology: do not invent it. Use the controlled English vocabulary. (rule 4)
- **NSE (Nortel):** Do not use jargon and idiomatic expressions. Use only terms that are in the NSE dictionaries. (rule 13)

These rules reflect and are consistent with the linguistic best-practice feature in Fig. 4.1 that the use of technical terms is limited in technical documents to help make the information understandable. They, therefore, provide an informed basis from which to develop the new non-manufacturer specific controlled language rule "Avoid using technical terms unless they are part of the controlled vocabulary.".

It has become evident in this and Chap. 2 that controlled languages aim at ensuring that words are spelled and used consistently in and across technical documents through control of spelling, part of speech and meaning. The fourteenth finding is that PACE and EE have one or more lexical rules that reflect, and are consistent with, either or both of the linguistic best-practice features from Fig. 4.1 that words are spelled and used consistently to help make the information in technical documents understandable. These rules are as follows:

- **PACE (Perkins):** Adhere to the PACE dictionary. (rule 8)
- **EE (Ericsson):** Do not use synonyms. (rule 1)
  Do not add prefixes or suffixes to EE words to make new words. (rule 3)

Three new non-manufacturer specific controlled language rules can be developed from these existing controlled language rules and the aforementioned best-practice features. The lexical rule "Use only words that are part of the controlled vocabulary." can be developed from PACE rule 8 and EE rule 3, the lexical rule "Use only the approved spelling, part(s) of speech and meaning(s) of words." can be developed from PACE rule 8 and the lexical rule "Do not add a new word to the controlled vocabulary if there is already a word with the same meaning." can be developed from EE rule 1.

The fifteenth and final finding is that the NSE rule "List events in sequence. Write events in the order in which they occur." (rule 4) reflects and is consistent with the organisational best-practice feature in Fig. 4.1 that the task-oriented chunks in technical documents are arranged in the order that the end users of the goods would generally be expected to need them to help make the information understandable. The new non-manufacturer specific controlled language rule addressing information structure "Arrange task-oriented chunks in the order that the end user is likely to need the information." can be developed from this NSE rule and the best-practice feature.

These fifteen findings suggest a strong link between the rules from existing controlled languages and the linguistic and organisational best-practice features of modern technical documents. The following two figures confirm this suggestion.

Controlled language (and manufacturer)	Rules that reflect one or more best-practice features	Total number (and percentage) of rules that reflect one or more best-practice features
CFE (Caterpillar)	1/2/3/4/5/6/8/9	8 of 10 rules (80%)
NSE (Nortel)	1/2/3/4/5/6/7/9/10/11/13/14/15	13 of 15 rules (87%)
BCE (Bull)	1/2/3/4/5/6/8/9	8 of 10 rules (80%)
PACE (Perkins)	1/2/4/7/8/9/10	7 of 10 rules (70%)
EE (Ericsson)	1/2/3/4/5/7/8/9/11/15/16/17	12 of 23 rules (52%)

Fig. 4.2 Breakdown of the identified rules

Fig. 4.2 provides a breakdown of the rules from each controlled language that were identified as reflecting one or more of the best-practice features.

The findings are particularly striking for CFE (Caterpillar), NSE (Nortel) and BCE (Bull) with at least 80% of their rules reflecting one or more of the best-practice features. In total, 48 or 71% of the 68 rules from the five controlled languages reflect one or more of the 22 best-practice features. Fig. 4.3 provides a breakdown of these best-practice

4 Analysing Existing Controlled Languages... 85

Finding	Best-practice feature	Classification
1	Uses short, simply structured sentences.	Linguistic
2	Does not use long strings of nouns and adjectives.	Linguistic
3	Limits each sentence to one piece of information or one instruction.	Organisational
4	Uses punctuation consistently.	Linguistic
5	Limits use of abbreviations and contractions.	Linguistic
6	Does not use colloquialisms, idioms and region-specific words.	Linguistic
7	Uses mainly positively worded sentences. Uses negatively worded sentences wherever possible in cautions and warnings.	Linguistic
8	Uses the active voice, particularly in sentences containing instructions.	Linguistic
9	Uses simple verb tenses and forms, in particular the simple present tense.	Linguistic
10	Avoids the gerund form.	Linguistic
11	Uses parallel language structures, particularly for task-oriented headings and sentences containing instructions.	Organisational
12	Does not use unnecessary words.	Linguistic
13	Does not use telegraphic language.	Linguistic
14	Spells words consistently.	Linguistic
15	Uses words consistently.	Linguistic
16	Arranges task-oriented chunks in the order that end users are likely to need the information.	Organisational

Fig. 4.3 Breakdown of the identified best-practice features

features. This reveals that the 48 rules reflect 16 or 73% of the best-practice features.

In conclusion, this strong link between the existing rules and the best-practice features has helped to identify 23 new non-manufacturer specific controlled language rules. These are as follows (in the order they were identified):

1. Use short sentences. (Limit sentences to no more than 25 words.)
2. Use simple sentence structures.
3. Limit word strings to no more than three nouns or adjectives.
4. Limit each sentence to one instruction.
5. Put two or more instructions in the same sentence if the end user is required to perform them at or almost at the same time.
6. Limit each sentence to one piece of information.
7. Use punctuation consistently. Always punctuate sentences containing similarly worded instructions in a similar way.
8. Avoid using abbreviations and contractions unless they are part of the controlled vocabulary.
9. Avoid using colloquialisms, idioms and region-specific words.
10. Use positively worded sentences wherever possible.
11. Use negatively worded sentences wherever possible in cautions and warnings to draw attention to their content.
12. Use the active voice wherever possible.
13. Always use the active voice in sentences containing instructions.
14. Use only the simple past, simple present, infinitive and simple future. Use the simple present wherever possible.
15. Avoid using gerunds unless they are part of the controlled vocabulary.
16. Use parallel language structures wherever possible, particularly for task-oriented headings and sentences containing instructions.
17. Avoid unnecessary words in sentences.
18. Do not omit articles (the, a, an), demonstrative adjectives (this, that, these, those) or relative pronouns (that, which, when) from sentences.
19. Avoid using technical terms unless they are part of the controlled vocabulary.

20. Use only words that are part of the controlled vocabulary.
21. Use only the approved spelling, part(s) of speech and meaning(s) of words.
22. Do not add new words to the controlled vocabulary unnecessarily. If you add a new word, provide an approved spelling, part(s) of speech and meaning(s) for it.
23. Arrange task-oriented chunks in the order that the end user is likely to need the information.

Summary

The findings from the analysis have shown that there is a strong link between the controlled language rules used by Caterpillar, Perkins, Ericsson, Nortel and Bull to help make the information in their technical documents understandable and the linguistic and organisational best-practice features identified from the literature as helping make the information in technical documents understandable. Specifically, 48 or 71% of the 68 rules reflect 16 or 73% of the 22 best-practice features. In addition, the findings from the analysis have helped identify 23 new non-manufacturer specific rules that will be used in the next chapter when developing a model controlled language rule set for technical documents.

Bibliography

Bernth, Arendse. 1997. EasyEnglish: A Tool for Improving Document Quality. In *Proceedings of the Fifth Conference on Applied Natural Language Processing*, 159–165. Seattle: Association for Computational Linguistics.
Lee, Arthur. 1993. Controlled English with and Without Machine Translation. In *Translating and the Computer 15*, ed. The Association for Information Management, 35–39. London: Aslib.
Nyberg, Eric H., Teruko Mitamura, and Willem-Olaf Huijsen. 2003. Controlled Language for Authoring and Translation. In *Computers and Translation*, ed. Harold Somers, 245–282. Amsterdam: John Benjamins Publishing Company.

O'Brien, Sharon. 2003. Controlling Controlled English: An Analysis of Several Controlled Language Rule Sets. In *Proceedings of EAMT-CLAW 2003*, 105–114. Dublin: Dublin City University.

Reuther, Ursula. 2003. Two in One-Can It Work? Readability and Translatability by Means of Controlled Language. In *Proceedings of EAMT/CLAW 2003*, 124–132. Dublin: EAMT/CLAW.

Schreurs, Dirk, and Geert Adriaens. 1992. Controlled English (CE): From COGRAM to ALCOGRAM. In *Computers and Writing: State of the Art*, ed. Patrik Holt and Noel Williams, 206–221. Oxford: Intellect Books.

Telefonaktiebolaget LM Ericsson. 1983. *Ericsson English Writer's Guide*. Stockholm: Telefonaktiebolaget LM Ericsson Standards Department.

ID
Developing a New Controlled Language for Technical Documents

Abstract The chapter first provides a model for developing a controlled language rule set that takes full account of all the identified linguistic and organisational best-practice features. The chapter then provides a model for developing a controlled vocabulary in which each word has an approved spelling, part(s) of speech and meaning(s).

Keywords Controlled language development • Controlled language models • Controlled language rule set • Controlled vocabulary

Introduction

Readers will recall from Chap. 2 that the controlled languages CFE, PACE, EE, NSE and BCE have a simplified set of rules and controlled vocabulary that together help make the information in technical documents understandable. This chapter will provide readers with models for developing a similar simplified set of rules and controlled vocabulary that they can use or adapt to develop a controlled language to help make the information in their technical documents understandable.

© The Author(s) 2017
S. Crabbe, *Controlling Language in Industry*,
DOI 10.1007/978-3-319-52745-1_5

The first part of the chapter provides a model for developing a controlled language rule set that builds on and extends the findings from the analysis described in the previous chapter to reflect and be consistent with all the linguistic and organisational best-practice features identified from the literature as helping make the information in technical documents understandable. The second part of the chapter provides a model for developing a controlled vocabulary in which each word has an approved spelling, part(s) of speech and meaning(s).

Model for Developing a Controlled Language Rule Set

It has previously been noted that a general lack of information about controlled languages persists to this day. Readers will therefore not be surprised to learn that there are, as far as the author is aware, no published (English language) guidelines or recommendations on controlled language rule set or controlled vocabulary sizes. The present section consequently starts by returning to the controlled languages analyses described in Chap. 4 to determine whether there is a standard size for controlled language rule sets. CFE, PACE and BCE have ten rules, which means that there is some consistency in the controlled languages analysed by the author. However, NSE has 15 rules and EE has 23 rules. There is thus no overall consistency.

The lack of consistency is even more pronounced in O'Brien's analysis, with just two of the controlled languages seeming to share the same rule set size. The word seeming is used here as, first, it was noted in Chap. 4 that descriptions of IBM's EasyEnglish suggest that it is not strictly a controlled language and, second, O'Brien's paper (2003) lists the number of rules in each of the analysed rule sets but not which number corresponds with which rule set. It is thus not possible to determine the number of rules in EasyEnglish. As a result, EasyEnglish and one of the controlled languages may share the same rule set size rather than two of the controlled languages. These findings strongly suggest that there is no standard size for controlled language rule sets.

In the absence of any guidelines, recommendations or standard size, the number of rules in the model controlled language rule set provided in this chapter has been determined on the basis of the number of rules necessary to ensure that each of 22 linguistic and organisational best-practice features identified from the literature as helping make the information in technical documents understandable is fully taken into account. The rules are as follows:

Section 1: Words

1.1 Use only words that are part of the controlled vocabulary.
1.2 Use only the approved spelling, part(s) of speech and meaning(s) of words.
1.3 Do not add new words to the controlled vocabulary unnecessarily. If you add a new word, provide an approved spelling, part(s) of speech and meaning(s) for it.
1.4 Avoid using abbreviations and contractions unless they are part of the controlled vocabulary.
1.5 Avoid using technical terms unless they are part of the controlled vocabulary.
1.6 Avoid using colloquialisms, idioms and region-specific words.
1.7 Limit word strings to no more than three nouns or adjectives.

Section 2: Verbs

2.1 Use the active voice wherever possible.
2.2 Always use the active voice in sentences containing instructions.
2.3 Use the imperative mood in sentences containing instructions.
2.4 Use only the simple past, simple present, infinitive and simple future. Use the simple present wherever possible.
2.5 Avoid using gerunds unless they are part of the controlled vocabulary.

Section 3: Sentences

3.1 Use short sentences. (Limit sentences to no more than 25 words.)
3.2 Use simple sentence structures.

3.3 Avoid unnecessary words in sentences.
3.4 Use positively worded sentences wherever possible. (See rule 3.5 for exception.)
3.5 Use negatively worded sentences wherever possible in cautions and warnings to draw attention to their content.
3.6 Use punctuation consistently. Always punctuate sentences containing similarly worded instructions in a similar way.
3.7 Do not omit articles (the, a, an), demonstrative adjectives (this, that, these, those) or relative pronouns (that, which, when) from sentences.

Section 4: Instructions

4.1 Limit each sentence to one instruction. (See rule 4.2 for exception.)
4.2 Put two or more instructions in the same sentence if the end user is required to perform them at or almost at the same time.
4.3 Use a numbered tabular structure for sentences containing consecutive instructions.
4.4 Arrange sentences containing instructions into task-oriented chunks.
4.5 Arrange task-oriented chunks in the order that the end user is likely to need the information.
4.6 Use task-oriented headings for task-oriented chunks.
4.7 Use parallel language structures wherever possible, particularly for task-oriented headings and sentences containing instructions.

Section 5: Information

5.1 Limit each sentence to one piece of information.
5.2 Arrange sentences containing information (not instructions) into short paragraphs of no more than nine sentences.
5.3 Limit each paragraph to one topic.

The rules can be classified as lexical (rules 1.1, 1.2, 1.3, 1.4, 1.5, 1.6, 1.7 and 4.6), grammatical (rules 2.1, 2.2, 2.3, 2.4, 2.5 and 3.6), syntactic (rules 3.1, 3.2 and 4.7), stylistic (rules 3.3, 3.4, 3.5 and 3.7) and as addressing information structure (rules 4.3, 4.4, 4.5 and 5.2) and information load (rules 4.1, 4.2, 5.1 and 5.3). This is broadly similar to CFE, PACE, EE, NSE and BCE. However, there is one difference that should be noted. None of the PACE rules have been classified as addressing information structure or information load; one of the CFE, EE and BCE rules has been classified as addressing information load and two of the NSE rules have been classified as addressing information structure or information load. However, eight of the model controlled language rules have been classified as addressing information structure or information load. This difference is explained by the fact that the model controlled language rule set is influenced by, but moves beyond CFE, PACE, EE, NSE and BCE, to reflect and be consistent with all the linguistic and organisational best-practice features identified from the literature.

The model controlled language rule set has been grouped into five sections of related rules. These groupings, which readers can choose whether or not to adopt, have been made with the aim of improving the understandability, and thus consistent interpretation, of the rules both individually and in combination with each other. The CFE, PACE, EE, NSE and BCE rules are not grouped into headed sections. However, the small rule set size of, in particular, CFE, PACE and BCE makes it difficult to group their rules into identifiable sections and negates the potential usefulness of doing this.

There are 19 rules in the first three sections of the model controlled language rule set entitled "Words", "Verbs" and "Sentences". Of these, 18 rules (rules 1.1, 1.2, 1.3, 1.4, 1.5, 1.6, 1.7, 2.1, 2.2, 2.4, 2.5, 3.1, 3.2, 3.3, 3.4, 3.5, 3.6 and 3.7) were developed from, and thus reflect and are consistent with, both linguistic best-practice features identified in Chap. 3 and existing controlled language rules analysed in Chap. 4. The remaining rule (rule 2.3) was developed from, and thus reflects and is consistent with, the linguistic best-practice feature identified in Chap. 3 that the imperative mood is used in sentences containing instructions to help make the information in technical documents understandable. For the convenience of readers who wish to use or adapt the seven rules grouped in the first section entitled "Words", Fig. 5.1 summarises their development.

Rule	Linguistic best-practice features developed from	Existing controlled language rules developed from
1.1 Use only words that are part of the controlled vocabulary.	Spells words consistently. **and** Uses words consistently.	PACE rule 8 and EE rules 1 and 3
1.2 Use only the approved spelling, part(s) of speech and meaning(s) of words.	Spells words consistently. **and** Uses words consistently.	PACE rule 8 and EE rules 1 and 3
1.3 Do not add new words to the controlled vocabulary unnecessarily. If you add a new word, provide an approved spelling, part(s) of speech and meaning(s) for it.	Spells words consistently. **and** Uses words consistently.	PACE rule 8 and EE rules 1 and 3
1.4 Avoid using abbreviations and contractions unless they are part of the controlled vocabulary.	Limits use of abbreviations and contractions.	CFE rule 8, EE rules 4 and 5, BCE rule 5 and NSE rule 11
1.5 Avoid using technical terms unless they are part of the controlled vocabulary.	Limits use of technical terms.	BCE rule 4 and NSE rule 13
1.6 Avoid using colloquialisms, idioms and region-specific words.	Does not use colloquialisms, idioms and region-specific words.	CFE rule 8, BCE rule 8, EE rule 4 and NSE rules 13 and 14
1.7 Limit word strings to no more than three nouns or adjectives.	Does not use long strings of nouns and adjectives.	CFE rule 4, PACE rule 9, EE rule 2 and NSE rule 10

Fig. 5.1 Development of rules 1.1–1.7

5 Developing a New Controlled Language for Technical...

Similarly, Fig. 5.2 summarises the development of the five rules grouped in the second section entitled "Verbs".

Rule	Linguistic best-practice features developed from	Existing controlled language rules developed from
2.1 Use the active voice wherever possible.	Uses the active voice, particularly in sentences containing instructions.	EE rule 8, NSE rule 6 and BCE rule 1
2.2 Always use the active voice in sentences containing instructions	Uses the active voice, particularly in sentences containing instructions.	EE rule 8, NSE rule 6 and BCE rule 1
2.3 Use the imperative mood in sentences containing instructions.	Uses the imperative mood, particularly in sentences containing instructions.	None
2.4 Use only the simple past, simple present, infinitive and simple future. Use the simple present wherever possible.	Uses simple verb tenses and forms, particularly the simple present tense.	NSE rule 9, EE rule 9, BCE rule 1 and CFE rule 6
2.5 Avoid using gerunds unless they are part of the controlled vocabulary.	Avoids the gerund form.	PACE rule 10, EE rule 16 and NSE rule 7

Fig. 5.2 Development of rules 2.1–2.5

Fig. 5.3 on the next page summarises the development of the seven rules grouped in the third section entitled "Sentences".

There are ten rules in the fourth and fifth sections of the model controlled language rule set entitled "Instructions" and "Information". Of these, five rules (rules 4.1, 4.2, 4.5, 4.7 and 5.1) were developed from, and thus reflect and are consistent with, both organisational best-practice features identified in Chap. 3 and existing controlled language rules analysed in Chap. 4.

The remaining rules (rules 4.3, 4.4, 4.6, 5.2 and 5.3) were developed from, and thus reflect and are consistent with, organisational

Rule	Linguistic best-practice features developed from	Existing controlled language rules developed from
3.1 Use short sentences. (Limit sentences to no more than 25 words.)	Uses short, simply structured sentences.	CFE rule 2, PACE rule 1, EE rule 15, NSE rule 2 and BCE rule 2
3.2 Use simple sentence structures.	Uses short, simply structured sentences.	BCE rule 5, NSE rule 3, CFE rule 2 and PACE rule 4
3.3 Avoid unnecessary words in sentences.	Does not use unnecessary words.	PACE rule 2 and EE rule 7
3.4 Use positively worded sentences wherever possible. (See rule 3.5 for exception.)	Uses mainly positively worded sentences.	CFE rule 1, BCE rule 1, NSE rule 1 and EE rule 11
3.5 Use negatively worded sentences wherever possible in cautions and warnings to draw attention to their content.	Uses negatively worded sentences wherever possible in cautions and warnings.	NSE rule 1 (sub-clause)
3.6 Use punctuation consistently. Always punctuate sentences containing similarly worded instructions in a similar way.	Uses punctuation consistently.	CFE rule 9, EE rule 17 and BCE rule 9
3.7 Do not omit articles (the, a, an), demonstrative adjectives (this, that, these, those) or relative pronouns (that, which, when) from sentences.	Does not use telegraphic language.	NSE rule 15 and PACE rule 7

Fig. 5.3 Development of rules 3.1–3.7

best-practice features identified in Chap. 3. Specifically, rule 4.3 was developed from the organisational best-practice features that a tabular structure is used for sentences, and a numbered tabular structure for sentences containing consecutive instructions, to help make the information in technical documents understandable. Rule 4.4 was developed from the organisational best-practice feature that sentences containing instructions are organised into task-oriented chunks in technical documents to help make the information understandable. Rule 4.6 was developed from the organisational best-practice feature that task-oriented headings are used for task-oriented chunks to help make the information in technical documents understandable. Lastly, rules 5.2 and 5.3 were developed from the organisational best-practice feature that sentences in technical documents containing information (rather than instructions) are chunked into short paragraphs of no more than nine sentences with one topic per paragraph to help make the information understandable.

For the convenience of readers who wish to use or adapt the seven rules grouped in the fourth section entitled "Instructions", Fig. 5.4 on the next page summarises their development.

Similarly, Fig. 5.5 on page 99 summarises the development of the three rules grouped in the fifth section entitled "Information".

It has been explained that the model controlled language rule set is grouped into sections of related rules with the aim to improve the understandability of the rules. It is also suggested that examples of the rules in practice are provided for the same reason. Fig. 5.6 on page 100 provides some examples to help readers in developing their own examples.

Model for Developing a Controlled Vocabulary

It was seen in Chap. 2 that CFE, PACE, EE, NSE and BCE all have a controlled vocabulary in which each word has an approved spelling, part(s) of speech and meaning(s). There are several methods that readers can use to develop a controlled vocabulary for their own technical documents.

The first method is to purchase a simplified dictionary and use the dictionary entries as the controlled vocabulary. Readers interested in this

Rule	Organisational best-practice features developed from	Existing controlled language rules developed from
4.1 Limit each sentence to one instruction. (See rule 4.2 for exception.)	Limits each sentence to one piece of information or one instruction.	CFE rule 3, EE rule 15, NSE rule 3 and BCE rule 4
4.2 Put two or more instructions in the same sentence if the end user is required to perform them at or almost at the same time.	Puts simultaneous, or nearly simultaneous, instructions in the same sentence.	CFE rule 3, EE rule 15, NSE rule 3 and BCE rule 4
4.3 Use a numbered tabular structure for sentences containing consecutive instructions.	Uses a tabular structure for sentences, particularly for sentences containing instructions. **and** Numbers sentences containing consecutive instructions.	None
4.4 Arrange sentences containing instructions into task-oriented chunks.	Arranges sentences containing instructions into task-oriented chunks.	None
4.5 Arrange task-oriented chunks in the order that the end user is likely to need the information.	Arranges task-oriented chunks in the order that end users are likely to need the information.	NSE rule 4
4.6 Use task-oriented headings for task-oriented chunks.	Uses task-oriented headings for task-oriented chunks.	None
4.7 Use parallel language structures wherever possible, particularly for task-oriented headings and sentences containing instructions.	Uses parallel language structures, particularly for task-oriented headings and sentences containing instructions.	CFE rule 5, NSE rule 5 and BCE rule 6

Fig. 5.4 Development of rules 4.1–4.7

5 Developing a New Controlled Language for Technical...

Rule	Organisational best-practice features developed from	Existing controlled language rules developed from
5.1 Limit each sentence to one piece of information.	Limits each sentence to one piece of information or one instruction.	CFE rule 3, EE rule 15, NSE rule 3 and BCE rule 4
5.2 Arrange sentences containing information (not instructions) into short paragraphs of no more than nine sentences.	Arranges sentences containing information (not instructions) into short paragraphs of no more than nine sentences with one topic per paragraph.	None
5.3 Limit each paragraph to one topic.	Arranges sentences containing information (not instructions) into short paragraphs of no more than nine sentences with one topic per paragraph.	None

Fig. 5.5 Development of rules 5.1–5.3

method may wish to look at, for example, the *McGraw-Hill Beginner's Dictionary of American English Usage* (Collin et al. 2002).

The second method is to purchase or obtain a commercially available controlled vocabulary and, if necessary, modify it. This method was used for EE, NSE and BCE. The University of Wales Institute of Science and Technology purchased the commercially available controlled language ILSAM from M. and E. White Consultants and modified its controlled vocabulary to create the EE controlled vocabulary. Nortel purchased the commercially available controlled language checker MAXit from Smart Communications and created the NSE controlled vocabulary from the MAXit dictionaries. Bull purchased the general dictionary which provided the BCE controlled vocabulary from Smart Communications.

Readers interested in this method may wish to look at, for example, the ASD-STE100 controlled vocabulary which has approximately 880

Controlling Language in Industry

[Approved examples are illustrated by the symbol ✓ and non-approved examples are indicated by the symbol ×.]

1.4 Avoid using abbreviations and contractions unless they are part of the controlled vocabulary.

Example ×

Don't disconnect the unit from the computer.

Example ✓

Do not disconnect the unit from the computer.

2.3 Use the imperative mood in sentences containing instructions.

Example ×

Sliding the HOLD switch to the ON position locks the controls.

Example ✓

Slide the HOLD switch to the ON position to lock the controls.

3.4 Use positively worded sentences wherever possible. (See rule 3.5 for exception.)

Example ×

Do not disconnect the unit from the computer before you click the "Safely Remove Hardware" icon on the computer.

Example ✓

1. Click the "Safely Remove Hardware" icon on the computer.

2. Disconnect the unit from the computer.

4.3 Use a numbered tabular structure for sentences containing consecutive instructions.

Example ×

To select a preset function, press the UP or DOWN button to select Preset. Press the MENU button to confirm the selection. Then, press the UP or DOWN button to select the desired preset function. Finally, press the MENU button to confirm the selection.

Example ✓

Select a preset function
1. Press the UP or DOWN button to select Preset.
2. Press the MENU button to confirm the selection.
3. Press the UP or DOWN button to select the desired preset function.
4. Press the MENU button to confirm the selection.

Fig. 5.6 Examples of some rules in practice

approved words. ASD-STE100 has been developed specifically to help make the information in aircraft maintenance documents understandable. However, the ASD-STE100 website (http://www.asd-ste100.org/faq.html) states that "Although STE was created to improve maintenance documentation, its principles can dramatically improve the reading quality of documentation in any industry" (2016).

The third method is to develop a controlled vocabulary from scratch. This method was used for CFE and PACE. Caterpillar developed the CFE controlled vocabulary from the most frequently recurring words in a broad sample of its maintenance and repair documents and Perkins developed the PACE controlled vocabulary from the most frequently recurring words in a broad sample of its installation documents.

For readers interested in the third method, this section provides a five-stage model for developing a controlled vocabulary from scratch based on the author's experience of controlled vocabulary development. Readers can use or adapt this model, shown below, to develop a controlled vocabulary for their own technical documents.

1. Compile a corpus of technical documents.
2. Generate a word frequency list of unique words in the corpus.
3. Identify multiple forms of nouns and verbs in the frequency list and remove all but one form of each.
4. Generate a concordance for each remaining word in the frequency list in order to identify its most frequently occurring part(s) of speech and meaning(s) in the corpus.
5. Identify any words in the frequency list that are used with the same part of speech and meaning in the corpus and remove all but one of them. The remaining most frequently occurring words, and their most frequently identified parts of speech and meanings from the concordances, form the controlled vocabulary.

The first suggested stage of the five-stage model is to compile a corpus of technical documents. There is general agreement in the literature on corpus development that there is no one minimum or optimum corpus size as it is dependent on the purpose of compiling the corpus (for example, Flowerdew 2004; Baker 2006; McEnery et al. 2006). The purpose of compiling the corpus in this case is the generation of a controlled vocabulary for technical documents. The content of the corpus should thus be a representative sample of all the types of technical documents that the controlled vocabulary is intended to be used for. It might thus include, for example, installation documents, service and maintenance documents and repair documents. It is recommended that readers save the corpus as a text file(s) so that corpus analysis software can be used to analyse the technical documents.

The second suggested stage of the model is to generate a word frequency list of every unique word in the corpus. This would be very difficult and time-consuming to do with pen and paper. It is thus recommended that readers use corpus analysis software to generate the word frequency list for their technical documents. These software packages are often free such as AntConc (http://www.laurenceanthony.net/software.html) or inexpensive such as MonoConc Pro (http://www.athel.com/mono.html). Fig. 5.7 shows a short extract from an example word frequency list to illustrate what one generally looks like.

Unique word	Raw frequency	Relative frequency (%)
turn	315	0.1688
sound	309	0.1656
format	307	0.1645
data	302	0.1618
usb	298	0.1597
connect	291	0.1560
software	251	0.1345

Fig. 5.7 Extract from an example word frequency list

The third suggested stage of the model is to identify any multiple noun forms (for example, keyboard and keyboards) and multiple verb forms (for example, download, downloads, downloading and downloaded) in the word frequency list and remove all but the singular and infinitive (dictionary) forms in each case.

This is because each noun is generally counted as one word and listed in the singular form only in controlled vocabularies, although it can be used in both the singular and plural forms. Similarly, each verb is generally counted as one word and listed in the infinitive (dictionary) form in controlled vocabularies, although it can be used in other approved forms. In the case of the model controlled language, these approved forms are listed in rule 2.4 as the simple past, simple present, infinitive and simple future tenses.

The fourth suggested stage is to generate a concordance for each remaining word in the frequency list in order to identify its most frequently occurring part(s) of speech and meaning(s) in the corpus. A concordance is simply a list of all the occurrences of a word in a corpus in the sentences in which they occur. This would again be very difficult and time-consuming to do with pen and paper. It is thus recommended that readers use the same corpus analysis software that they used to generate the word frequency list in the second stage. The following shows a short extract from an example concordance for the unique word "field" to illustrate what one generally looks like.

> … enter text in the search **field**. Enter text in the search …
> … screen showing the search **field** at the top and information …
> … of the screen is a search **field**. Below the search field is …
> … you enter text in the search **field**, contacts with information …
> … button. Beside the search **field** from left to right is the …
> … a keyword in the search **field**. Search results appear …
> … To the left of the search **field** is the Delete button. …
> … To the right of the search **field**, contacts with information …

The concordance lines reveal that "field" occurs as part of the lexical chunk "search field" and is used as one part of speech and with one meaning in the corpus, that is, as a noun meaning an on-screen area in which text can be entered. This means that "field" would be approved in the controlled vocabulary as a noun that means an on-screen area in which text can be entered.

This stage is likely to be the most time-consuming. However, it is important that it is carried out accurately and completely to ensure that all the approved spellings, parts of speech and meanings accurately reflect the corpus content.

The final suggested stage of the five-stage model is to identify any words in the frequency list that are used with the same part of speech and meaning in the corpus and remove the least-frequently occurring word. For example, if the verbs "select" and "choose" or "save" and "store" are used with the same meaning, then the least-frequently occurring verb in each verb pair should be removed. The remaining words in the frequency list, and their identified spelling, part(s) of speech and meaning(s), will become the controlled vocabulary for the new controlled language.

It should be made clear at this point that different readers are likely to have a different number of words remaining. This is something to be expected as the number of remaining words will depend on, first, the total number of unique words in the corpus in the first place and, second, the number of nouns, verbs and synonyms that are deleted. Furthermore, readers will recall from the previous section of this chapter that there are no published (English language) guidelines or recommendations on controlled vocabulary sizes.

In concluding this section, two points must be made. First, a controlled vocabulary is not static. It will need to be revised or extended if, for example, a manufacturing company's line of goods changes or new goods are added. One potential, simple method of doing this is to allow the controlled vocabulary to be extended by new user-approved specialised technical terms.

Second, a controlled vocabulary is not intended to be used in isolation but rather in accordance with a controlled language rule set to help make the information in technical documents understandable.

Summary

This chapter has provided readers with models for developing a controlled language (a simplified set of rules and controlled vocabulary) for their own manufacturing company. The final stage prior to implementing a newly developed controlled language is to conduct a pilot trial. The next, concluding, chapter will thus provide readers with a suggested model for conducting a pilot trial of a controlled language before implementing it company-wide.

Bibliography

Baker, Paul. 2006. *Using Corpora in Discourse Analysis*. London: Continuum International Publishing Group.

Collin, Peter, Miriam Lowi, and Carol Weiland. 2002. *McGraw-Hill Beginner's Dictionary of American English Usage*. 2nd ed. London: McGraw-Hill.

Flowerdew, Lynne. 2004. The Argument for Using English Specialized Corpora to Understand Academic and Professional Language. In *Discourse in the Professions: Perspectives from Corpus Linguistics*, ed. Ulla Connor and Thomas A. Upton, 11–33. Amsterdam: John Benjamins Publishing Company.

McEnery, Tony, Richard Xiao, and Yukio Tono. 2006. *Corpus-Based Language Studies*. London: Routledge.

O'Brien, Sharon. 2003. Controlling Controlled English: An Analysis of Several Controlled Language Rule Sets. In *Proceedings of EAMT-CLAW 2003*, 105–114. Dublin: Dublin City University.

6

Trialling a New Controlled Language for Technical Documents

Abstract The concluding chapter provides suggestions on how to conduct a pilot trial of a new controlled language before deciding whether or not to implement it company-wide. The objective of the pilot trial is to assess the effect, if any, of the application of the controlled language on how understandable the information in technical document extracts is. The chapter finishes with a final note on the book's aim and content.

Keywords Controlled language trialling • Reading tests • Understandable technical documents

Introduction

Readers will not be surprised to learn that there are, as far as the author is aware, no published (English language) guidelines on how to conduct a pilot trial of a new controlled language. There are also a very limited number of publicly available studies assessing the effectiveness of individual controlled languages on technical documents. The chapter thus

© The Author(s) 2017
S. Crabbe, *Controlling Language in Industry*,
DOI 10.1007/978-3-319-52745-1_6

first provides readers with a suggested model for conducting a pilot trial of a controlled language before deciding whether or not to implement it company-wide. The objective of the pilot trial is to assess whether the revision of technical document extracts for compliance with a controlled language affects how understandable the information in the technical document extracts is for the target readers. There is thus no requirement to have access to or to use a usability laboratory. Readers interested in conducting a full-scale usability test are referred to Byrne (2006). The chapter, and the book, then concludes with a final note on the book's aim and content.

Model for Conducting a Pilot Trial

This section provides a six-stage model for conducting a pilot trial of a new controlled language. This is based on the author's experience of controlled language trialling, and a study conducted by Shubert, Holmback, Spyridakis and Coney in 1995 and described in two papers "Testing the Comprehensibility of Simplified English: An Analysis of Airplane Procedure Documents" (1995a) and "The Comprehensibility of Simplified English in Procedures" (1995b). As with the model for developing a controlled vocabulary described in Chap. 5, it is hoped that readers reproduce or adapt the model, shown here, for their own use.

1. Select extracts from two technical documents.
2. Revise the extracts for compliance with the new controlled language.
3. Design questions to measure the understandability of the original and revised extracts.
4. Recruit the trial participants.
5. Have the trial participants complete the question set for the assigned technical document extract.
6. Analyse the findings.

The first suggested stage of the model is to select the material for the pilot trial. This involves selecting suitable technical documents and then selecting suitable extracts from the technical documents. It is suggested that two technical documents of a similar type and level of understandability are selected in order to strengthen the findings. With regard to technical document type, this could mean, for example, two installation documents, two service and maintenance documents or two repair documents. In the case of Shubert, Holmback, Spyridakis and Coney, the selected documents were service and maintenance documents.

With regard to understandability, it was revealed in Chap. 3 that many tests have been developed to estimate technical document understandability. Readers who wish to check the understandability of their technical documents are encouraged to use the Automated Readability Index and/or Flesch-Kincaid Grade Level Formula. This is because, first, there are many free online versions of both tests available. Second, both tests have been developed specifically to estimate English language technical document understandability. Third, they both estimate understandability in terms of US grade levels, thus making a comparison of their results possible. Finally, they use different linguistic features and mathematical formulas, thus providing two independent estimates of understandability. The Automated Readability Index is calculated as: GL = (4.71 × average number of letters per word) + (0.5 × average number of words per sentence) − 21.43 where GL is the US grade level needed to understand the technical document. The Flesch-Kincaid Grade Level Formula is calculated as: GL = (0.39 × average number of words per sentence) + (11.8 × average number of syllables per word) − 15.59 where GL is again the US grade level needed to understand the technical document.

It is suggested that readers next select an extract of a similar length and type from each technical document. This might, for example, mean a 750-word extract containing procedural instructions, a 750-word extract containing cautions and warnings or a 750-word extract containing non-procedural information. In the case of Shubert, Holmback, Spyridakis and Coney, the selected extracts were 686 and 846 words and contained procedural instructions.

The choice of extract length and type is up to the reader, but it is important that both extracts are a similar length and type in order to strengthen the findings.

The second suggested stage of the model is to revise the extracts for compliance with both the simplified set of rules and controlled vocabulary from the new controlled language. In doing this, four technical document extracts are created for the pilot trial: the first original extract, the revision of the first original extract for compliance with the new controlled language, the second original extract and the revision of the second original extract for compliance with the new controlled language.

Readers should not be surprised if either or both of their revised extracts are shorter (in terms of word count) than the corresponding original extracts. This is likely due, for example, to the elimination of inconsistencies in the usage of words in the original extracts through compliance with rule 1.1 "Use only words that are part of the controlled vocabulary." and rule 1.2 "Use only the approved spelling, part(s) of speech and meaning(s) of words.".

The third suggested stage of the model is to draw up two sets of questions: one set to assess the understandability of the information in the first original extract and the revision of the first original extract and one set to assess the understandability of the information in the second original extract and the revision of the second original extract. The content of each set of questions will depend on the content of each technical document extract. However, the overall focus of each set of questions will likely be similar to that of Shubert, Holmback, Spyridakis and Coney who state that "We attempted to focus on those parts of the procedures that had important SE-related differences in language versus the non-SE versions of the documents" (1995b: 357), where the acronym refers to the controlled language Simplified English.

Readers should also try, as far as possible, to keep the number of questions, and the balance of question types (for example, short-answer questions, true-false questions and multiple-choice questions), similar in each set of questions in order to strengthen their findings. In the case of Shubert, Holmback, Spyridakis and Coney, the number of questions was the same in each set of questions, but the balance of question types was slightly different because there were differences in the content of each technical document extract. Fig. 6.1 shows the balance of question types in the study conducted by Shubert, Holmback, Spyridakis and Coney.

	Number of questions		
	Short-answer	Multiple-choice	True-false
Set of questions for extract A	10	5	5
Set of questions for extract B	14	3	3

Fig. 6.1 Balance of question types

Finally, it is suggested that the questions in each set are arranged in a different order from the information in the technical document extracts. This is because researchers (for example, Griffiths 2001; Pringle and O'Keefe 2003; Byrne 2006; Kimball and Hawkins 2008) agree that technical documents are often read selectively for reference purposes rather than from cover to cover. The arrangement of the questions in a non-sequential order aims to approximate this use of technical documents.

The fourth suggested stage of the model is to select participants for the pilot trial. All things being equal, the larger the number of participants, the more reliable the findings will be. However, for practical, financial and organisational reasons, it may not be possible to recruit a large number of participants for a pilot trial. Nevertheless, the following is recommended. First, to the extent that it is possible, the recruited participants should be representative of the target readers of the technical documents. For example, in the case of technical documents for target readers with mixed English language and/or literacy skills, the recruited participants should include readers with mixed English language and/or literacy skills.

Second, to the extent that it is possible, the participants should not have used the engine, machine or equipment in the technical document extract. This is to ensure that the participants' answers are not influenced by prior use of the engine, machine or equipment.

Third, to the extent that it is possible, an equal number of participants should be randomly assigned to each of the four technical document extracts in order to strengthen the findings. Shubert, Holmback, Spyridakis and Coney do not describe their selection criteria; thus, it is not known if comparable criteria were employed in their study.

The fourth suggested stage of the model is to have the participants complete the set of questions for their assigned technical document extract. There is no set method for doing this. However, some suggestions include, first, having the participants complete the set of questions in as quiet and comfortable a room as possible to minimise movement and noise distractions and, second, asking the participants to not discuss the questions with any future participants to minimise sharing of answers.

The final suggested stage of the model is to mark the test questions for each of the four technical document extracts to see if, and by how much, the number of correct answers was higher for the participants randomly assigned either the revision of the first original extract or the revision of the second original extract when compared with the participants randomly assigned either the first original extract or the second original extract.

A higher number of correct answers, particularly for both sets of questions, suggest that the use of the controlled language helped make the information more understandable for the target readers. In this case, readers may decide to proceed with implementing the controlled language company-wide. A decrease in correct answers, particularly for both sets of questions, alternatively suggests that use of the controlled language did not help make the information more understandable for the target readers. In this case, readers may wish to conduct a second pilot trial with different technical document extracts, sets of questions and/or participants to further assess the controlled language before deciding whether or not to implement it company-wide.

Final Note

At the First International Workshop on Controlled Language Applications in 1996, Goyvaerts argued that "Industry does not need Shakespeare or Chaucer, industry needs clear, concise communicative writing—in one word Controlled Language" (cited by Spaggiari and Cardey 2010: 394). However, 20 years later, when the author started writing this book, there was still no complete book on the market that addressed controlled languages. The aim of this book has thus been to introduce readers to the purpose, use, development and evaluation of controlled languages so as to provide them with a clear understanding of existing controlled languages and models which they can use or adapt to develop and trial a new controlled language for their own technical documents. It is hoped that the book has achieved this aim and, in doing so, will encourage readers to consider developing controlled languages for their manufacturing companies.

Bibliography

Byrne, Jody. 2006. *Technical Translation: Usability Strategies for Translating*. Dordrecht: Springer.
Griffiths, Dave. 2001. Design for Usability—No One Reads Manuals. In *The ISTC Handbook of Professional Communication and Information Design*, ed. The Institute of Scientific and Technical Communicators, 1–12. Peterborough: The Institute of Scientific and Technical Communicators.
Kimball, Miles A., and Ann R. Hawkins. 2008. *Document Design: A Guide for Technical Communicators*. Boston: Bedford/St. Martin's.
Pringle, Alan S., and Sarah S. O'Keefe. 2003. *Technical Writing 101: A Real-World Guide to Planning and Writing Technical Documentation*. 2nd ed. Dallas: Scriptorium Press.
Shubert, Serena, Heather Holmback, Jan Spyridakis, and Mary Coney. 1995a. Testing the Comprehensibility of Simplified English: An Analysis of Airplane Procedure Documents. In *IEEE International Professional Communication Conference: IPCC 95 Proceedings: Smooth Sailing to the Future*, 171–173. Savannah: IEEE Press.
———. 1995b. The Comprehensibility of Simplified English in Procedures. *Journal of Technical Writing and Communication* 25(4): 347–369.
Spaggiari, Laurent, and Sylviane Cardey. 2010. A System to Control Language for Oral Communication. In *Advances in Natural Language Processing: 7th International Conference on NLP, IceTAL 2010, Reykjavik, Iceland, August 2010, Proceedings*, ed. Hrafn Loftsson, Eirikur Rögnvaldsson, and Sigrun Helgadóttir, 393–400. Berlin: Springer.

Index

A

abbreviations, 26, 30, 35, 38, 40, 41, 51, 55, 74, 78, 79, 85, 86, 91, 94, 100

active voice, 7, 13–15, 17, 35, 38, 51, 52, 72–4, 80, 85, 86, 91, 95

adjective strings, 29, 33, 35, 51, 55, 74, 76, 85, 86, 91, 94

Adriaens, Geert, 37, 70–3, 76

American industrial revolution, 12–16, 24

ASD-STE100, 99, 101

Automated Readability Index, 57, 109

B

Basic English, 2, 18, 24–30, 32, 33

BCE. *See* Bull Controlled English (BCE)

benefits of understandable technical documents, 2, 51, 60–1

British industrial revolution, 9–12, 24

Bull Controlled English (BCE), 39–42, 73–5, 77–84, 89, 90, 93–9

C

Caterpillar Fundamental English (CFE), 28–33, 35–7, 39, 41, 42, 69, 73, 75–82, 84, 89, 90, 93–9, 101

Caterpillar Technical English (CTE), 31, 34

CFE. *See* Caterpillar Fundamental English (CFE)

Index

Chambers, Ephraim
   Cyclopedia, 7, 8
Chaucer, Geoffrey, 112
   A Treatise on the Astrolabe, 5–7, 12, 58
chunking principle, 59
colloquialisms, 26, 30, 40, 41, 51, 55, 74, 78, 79, 85, 86, 91, 94
concise language, 53
Coney, Mary, 108–10, 112
consistent spelling, 51, 54, 71, 74, 83, 85, 94
consistent word use, 51, 55, 71, 73, 74, 83, 85, 94
contractions, 30, 35, 41, 51, 55, 74, 78, 79, 85, 86, 91, 94, 100
controlled language benefits, 2, 25, 42–3
controlled language drawbacks, 2, 25, 42–3
controlled language pilot trial, 2, 3, 105, 107–12
controlled language rule set model, 2, 3, 70, 74, 87, 89–97, 104
controlled vocabulary, 1, 3, 24, 25, 27–30, 32, 33, 35–7, 39, 40, 42, 71, 72, 79, 81, 83, 86, 97, 99–104, 108, 110
controlled vocabulary model, 3, 89, 90, 97–104
conversational expressions, 35, 78, 79
CTE. *See* Caterpillar Technical English (CTE)

D
dual-oriented controlled language, 32

E
Ericsson English (EE), 34–7, 39, 73, 75–84, 89, 90, 93–9
Evans, Oliver, 13
   The Young Mill-Wright and Millers Guide, 12, 13

F
Flesch-Kincaid Grade Level Formula, 57, 109
Ford, Henry, 14, 25
Ford Model T, 14–17

G
German industrial revolution, 16–17
gerund form, 38, 51, 54, 72–4, 81, 85, 91, 95

H
Harris, John
   Lexicon Technicum, 7, 8
Holmback, Heather, 108–10, 112
human-oriented controlled language, 29

I
idioms, 35, 38, 51, 55, 74, 78, 79, 83, 85, 86, 91, 94
imperative mood, 7, 13–15, 17, 51, 52, 74, 91, 93, 95, 100
instructional sentences, 7, 13–15, 17, 52, 57–9, 80

Index

L
linguistic best-practice features, 2, 3, 7, 13, 51–7, 60, 62, 70–6, 78–85, 87, 90, 91, 93–6
literacy, 10, 11, 52, 53, 56, 59, 111

N
negative wording, 35, 37, 38, 51, 53, 74, 79, 80, 85, 86, 92, 96
Nortel Standard English (NSE), 37–9, 73, 75–84, 89, 90, 93–9
noun strings, 29, 33, 35, 38, 51, 55, 71–4, 76, 85, 86, 91, 94

O
O'Brien, Sharon, 29, 32, 70–3, 75, 76, 80–2, 90
Ogden, Charles, 18, 24, 25, 27, 28
organisational best-practice features, 2, 3, 7, 13, 50, 57–60, 62, 70, 74, 77, 82, 84, 87, 90, 91, 93, 95, 97–9

P
PACE. *See* Perkins Approved Clear English (PACE)
paragraphs
 information load of paragraphs, 57, 59, 74, 92, 97, 99
 paragraph length, 57, 59, 74, 92, 97, 99
parallel language, 7, 17, 26, 38, 40, 41, 57, 58, 74, 81, 82, 85, 92, 98

Perkins Approved Clear English (PACE), 31–7, 39, 73, 75, 76, 81–4, 89, 90, 93–7, 101
positive wording, 26, 29, 37, 40, 41, 51, 53, 74, 79, 80, 85, 86, 92, 96, 100
pre-industrial Britain, 4–9
punctuation, 26, 30, 36, 40, 41, 51, 54, 74, 78, 85, 86, 92, 96

R
region-specific words, 38, 51, 55, 74, 79, 86, 91, 94

S
Schreurs, Dirk, 37, 70–3, 76
sentences
 information load of sentences, 26, 30, 36, 38–40, 77, 93
 sentence chunking, 59, 60, 74, 92, 97, 98
 sentence length, 34, 35, 37, 40, 41, 51–2, 70, 72–7, 85, 86, 91, 96
 sentence numbering, 13, 14, 17, 57–9, 74, 92, 97, 98, 100
 sentence structuring, 7, 15, 17, 29, 38–41, 51–2, 57–9, 74–6, 81–2, 85, 86, 91, 92, 96–8, 100
Shubert, Serena, 27, 37, 108–10, 112
Spyridakis, Jan, 108–10, 112

T

tabular structuring, 57–9, 74, 92, 97, 98, 100
task orientation, 6, 7, 12–15, 17, 57–60, 74, 82, 84–7, 92, 97, 98
technical terms, 8, 28, 30, 33, 42, 51, 54, 55, 74, 83, 86, 91, 94, 104
telegraphic language, 51, 53, 72–4, 82, 85, 96

U

Urquhart, Thomas

Logopandecteision, 24

V

verb tenses, 35, 38, 51, 54, 71, 74, 80, 81, 85, 91, 95

W

warnings and cautions, 36, 37, 51, 53, 61, 74, 80, 85, 86, 92, 96, 109
Wilson Sewing Machine Company, 13, 14

The manufacturer's authorised representative in the EU is Springer Nature Customer Service Centre GmbH, Europaplatz 3, 69115 Heidelberg, Germany. If you have any concerns regarding our products, please contact ProductSafety@springernature.com

Printed and bound by CPI Group (UK) Ltd, Croydon, CR0 4YY
23/03/2026
02076402-0002